THE NAIL

Other titles in the GEM Monograph series

Dupuytren's Disease
Edited by John T. Hueston and Raoul Tubiana
Monograph No 1 1974 Second edition 176 pages illustrated

Traumatic Nerve Lesions of the Upper Limb
Edited by J. Michon and Erik Moberg
Monograph No 2 1974 124 pages illustrated

Mutilating Injuries of the Hand
Edited by D. A. Campbell Reid and the late J. Gosset
Monograph No 3 1979 144 pages illustrated

Tendon Surgery of the Hand
Edited by Claude Verdan
Monograph No 4 1979 196 pages illustrated

General Series Editor: R. Tubiana

THE NAIL

Edited by Maurice Pierre

IN COLLABORATION WITH

G. Achten, P. Banzet, R. Baran, B. Barfod,   R. Malek, J. Michon, E. Moberg,
H. Bureau, J. P. Delagoutte, C. Dufourmentel,   W. A. Morrison, R. Mouly, J. Pillet, A. Thion,
M. Iselin, J. P. Jouglard, C. R. McCash,   H. Tramier, C. Verdan

FIRST ENGLISH EDITION

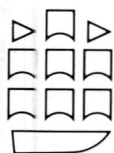

CHURCHILL LIVINGSTONE
EDINBURGH LONDON MELBOURNE AND NEW YORK 1981

CHURCHILL LIVINGSTONE
Medical Division of Longman Group Limited

Distributed in the United States of America by Churchill Livingstone
Inc., 19 West 44th Street, New York, N.Y. 10036, and by associated
companies, branches and representatives throughout the world.

First edition published in French under the title L'ongle
© Expansion Scientifique Française 1978

First edition in English based on French edition
© Longman Group Limited 1981

ISBN 0 443 02102 3

British Library Cataloguing in Publication Data
The Nail. – (GEM monographs)
 1. Nails (Anatomy) – Diseases
 I. Pierre, Maurice II. L'ongle. *English*
 616.5'47   RL165

Library of Congress Catalog Card Number 81–9935

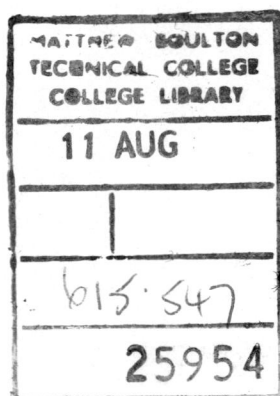

Printed and bound in Great Britain by
William Clowes (Beccles) Limited, Beccles and London

Preface

Studies of disorders of the hand have generally given insufficient attention to the nail, despite its functional importance and aesthetic value. It protects the dorsal surface of the fingertip, enabling us to grasp small objects and to scratch and, together with the digital pulp, it contributes to the acute sensitivity of the fingertip. In addition to these functional aspects, it plays a prominent role in aesthetics, being continually exposed to view. To a woman, in particular, a well-shaped nail, delicately varnished, is an important cosmetic attribute, while an absent or deformed nail is a justifiable source of concern.

To the dermatologist, for whom the nail is an appendage of the skin, and to the surgeon, the pathology of the nail poses problems of diagnosis and treatment. Dermatological treatment is frequently prolonged, while surgery, although more rapid, can give rise to its own range of complications, which may be difficult to treat. The objective of the surgeon is to preserve or improve the function and appearance of nails, which have been damaged by disease, injury, or congenital malformation. Conservation is not always possible in the case of tumours, malformations and severe burns.

The purpose of this monograph by the Groupe d'Étude de la Main is to assemble the different aspects of nail pathology, both from the medical and surgical aspects. It represents the accumulated effort and experience of those who have been interested in this frequently neglected field.

M. Pierre

Contributors

G. *ACHTEN*, Clinique de dermatologie et de syphiligraphie, Hôpital Universitaire Saint-Pierre, Université libre de Bruxelles, 322 Rue Haute, Bruxelles, Belgium

P. *BANZET*, Service de Chirurgie Plastique Reconstructive, Hôpital Saint-Louis, 2 Place du Docteur A. Fournier, 75475 Paris Cedex 10, France

R. *BARAN*, Centre Hospitalier, 06407 Cannes, France

B. *BARFOD*, Department of Orthopaedic Surgery, Aarhus Hospital, Aarhus, Denmark

H. *BUREAU*, Hôpital de la Timone, Boulevard d'Alès, 13005 Marseille, France

J. P. *DELAGOUTTE*, Institut de Réadaptation, Rue Lionnois, 54000 Nancy, France

C. *DUFOURMENTEL*, Chirurgie Plastique Reconstructive, Hôpital Saint-Louis, 2 Place du Docteur A. Fournier, 75475 Paris Cedex 10, France

M. *ISELIN*, 1 Rue Auguste-Vacquerie, 75116 Paris, France

J. P. *JOUGLARD*, Service de Chirurgie Plastique, Hôpital de la Conception, 144 Rue Saint-Pierre, 13005 Marseille, France

C. R. *McCASH*, London, Great Britain

R. *MALEK*, Service de Chirurgie Infantile, Hôpital St-Vincent de Paul, 74 Avenue Denfert-Rochereau, 75674 Paris Cedex 14, France

J. *MICHON*, Hôpital Jeanne-d'Arc, Dommartin-lès-Toul, 54000 Toul, France

E. *MOBERG*, Hand Surgery Clinic, Sahlgrenska Sjukhuset, Gothenburg, Sweden

W. A. *MORRISON*, St Vincent Hospital, Melbourne, Australia

R. *MOULY*, Service de Chirurgie Plastique Reconstructive, Hôpital Saint-Louis, 2 Place du Docteur A. Fournier, 75475 Paris Cedex 10, France

M. *PIERRE*, 38 Boulevard Périer, 13008 Marseille, France

J. *PILLET*, Centre de Prothèse Plastique et Restauratrice, Hôpital Saint-Antoine, 184 Rue du Fg-St Antoine, 75571 Paris Cedex 12, France

A. *THION*, Centre Borely, 2 Avenue du Parc-Borely, 13008 Marseille, France

H. *TRAMIER*, 41 Rue St. Jacques, 13006 Marseille, France

C. *VERDAN*, Policlinique Chirurgicale Universitaire, 9 Avenue de la Gare, 1003 Lausanne, Switzerland

Contents

1. **Histopathology of the nail** *G. Achten*   1
2. **Nail malformations** *R. Malek*   15
3. **Principal modifications of the normal form of the nail** *R. Baran*   19
4. **Modifications of the nail surface** *R. Baran*   26
5. **Modifications of colour: chromonychias or dyschromias** *R. Baran*   30
6. **Onychia and paronychia of mycotic, microbial and parasitic origin** *R. Baran*   39
7. **The nail in dermatological disease** *R. Baran*   46
8. **Nail tumours** *R. Baran*   54
9. **Nail changes in general pathology** *R. Baran*   61
10. **Nail changes in principal genetic diseases** *R. Baran*   65
11. **Melanotic tumours in the nail area** *P. Banzet, R. Mouly and C. Dufourmentel*   74
12. **Glomus tumours** *H. Bureau, J. P. Jouglard, A. Thion, H. Tramier and M. Pierre*   76
13. **The nail in radiodermatitis** *M. Pierre*   80
14. **Crush injuries of the digital extremities** *J. Michon and J. P. Delagoutte*   81
15. **Avulsion injuries of the nail** *M. Iselin*   83
16. **Nail prostheses** *C. Dufourmentel*   85
17. **Prostheses in amputations of the fingertips** *J. Pillet*   90
18. **Plastic surgery and the claw nail** *C. Verdan*   93
19. **Mucous pseudocysts of the finger** *E. Moberg*   102
20. **Free nail grafting** *C. R. McCash*   104
21. **Reconstruction of the finger-nail by microvascular transfer from the toes** *W. A. Morrison*   108
22. **Reconstruction of the nail fold** *B. Barfod*   114
Index   117

GROUPE D'ETUDE DE LA MAIN (G.E.M.)

List of Members

U.K.
D. M. Brooks (London) G. Fisk (Essex) S. H. Harrison (London) J. I. P. James (Edinburgh) D. Lamb (Edinburgh) F. Nicolle (London) R. G. Pulvertaft (Derby) R. H. C. Robins (Cornwall) H. J. Seddon (London) H. G. Stack (Essex) C. B. Wynn Parry (London)

U.S.A.
A. J. Barsky (New York) R. Beasley (New York) J. Bell (Chicago) J. Boswick (Colorado) J. Boyes (California) P. Brand (California) S. Brown (California) E. Clark (California) R. Curtis (Maryland) A. E. Flatt (Iowa) J. Hunter (Philadelphia) E. Kaplan (New York) C. H. Lane (California) W. J. Littler (New York) J. W. Madden (Arizona) L. Milford (Tennessee) E. Nalebuff (Massachusetts) M. Spinner (New York) A. B. Swanson (Michigan)

Australia
M. Fahrer (Queensland) J. Hueston (Victoria) W. Morrison (Melbourne)

France
P. C. Achach (Paris) Y. Allieu (Montpellier) J. Y. Alnot (Paris) J. Aubriot (Caen) P. Banzet (Paris) J. Baudet (Bordeaux) S. Baux (Paris) J. Beres (Paris) R. Bobichon (Grenoble) J. Body (Chaumont) M. Bombart (Villeneuve St Georges) M. Bonnel (Montpellier) J. Bonvallet (Paris) A. Borit (St Maur les Fosses) P. Bourrel (Marseille) J. L. Brouet (Toulon) P. de Butler (Amiens) H. Bureau (Marseille) J. Carayon (Marseille) A. Chancholle (Toulouse) P. Colson (Lyon) J. J. Comtet (Lyon) J. C. Dardour (Paris) P. Dautry (Paris) J. Dubousset (Clamart) J. L. Ducourtioux (Paris) C. Dufourmentel (Paris) J. Duparc (Paris) J. S. Elbaz (Paris) P. Esteve (Neuilly) G. Foucher (Strasbourg) Fourrier (Clermont-Ferrand) M. Gangolphe (Ste Foy les Lyon) R. Gay (Toulouse) A. Gilbert (Paris) J. Glicenstein (Paris) A. Goumain (Bordeaux) J. Gournet (Reims) J. Greco (Tours) C. Hamonet (Paris) S. Hauttier (Paris) F. Iselin (Paris) M. Iselin (Paris) M. Jandeaux (Vesoul) J. P. Jouglard (Aubagne) A. Julliard (Neuilly) A. Kapandji (Longjumeau) M. Kerboul (Paris) N. Kuhlmann (Beauvais) J. P. Lalardrie (Paris) F. Langlais (Paris) Lemerle (Paris) A. Lemoine (Paris) Le Quang (Paris) J. Lerique (Paris) J. Levans (Nanterre) J. Lignon (Nantes) R. Lisfranc (Neuilly les Toul) R. Malek (Paris) M. Mansat (Toulouse) J. P. May (Paris) C. Menkes (Paris) M. Merle (Dommartin les Toul) R. Merle d'Aubigne (Achéres) J. P. Meyreuis (Toulon) J. Michon (Dommartin les Toul) C. Moitrel (Bois Guillaume) D. Morel-Fatio (Paris) R. Mouly (Paris) R. Naett (Strasbourg) C. Nicoletis (Paris) J. A. Noirclerc (Collonges au Mont D'Or) P. Oger (Garches) P. Petit (Paris) M. Pierre (Marseille) J. Pillet (Paris) J. G. Pous (Montpellier) P. Rabischong (Montpellier) J. P. Razemon (Lille) J. P. Rengeval (Paris) J. Roulet (Lyon) C. Roux (Montrouge) P. Saffer (Neuilly) T. Saucier (Grenoble) A. Sedel (Jouy en Josas) R. Souquet (Toulouse) R. Thévenin (Rouen) J. M. Thomine (Rouen) H. Tramier (Marseille) R. Tubiana (Paris) P. Valentin (Clermont-Ferrand) J. M. Vaillant (Neuilly) C. Valette (Limoges) B. Valtin (Champigny) P. Vichard (Besancon) R. Vilain (Saint Cloud)

Austria
A. Berger (Vienna) E. Trojan (Vienna) H. Millesi (Vienna)

Belgium
De Conninck (Brussels) H. Evrard (Jamioulx) H. de Frenne (Waregem) P. Van Wetter (Brussels)

Finland
K. Vainio (Heinola)

Germany
D. Buck-Gramko (Hamburg) Haimovici (Bremen) L. Mannerfelt (Villingen am Schwarzwald)

Holland
J. Bloem (Heemstede) J. F. Landsmeer (Leiden) J. Van Der Meulen (Rotterdam)

Italy
P. Bedeschi (Modena) G. Brunelli (Brescia) A. Gensini (Rome) R. Mantero (Savona) E. Morelli (Cerro Maggiore) V. Salvi (Turin)

Luxemburg
J. Y. de la Caffiniére

Spain
F. Enriquez de Salamanca (Madrid) Quintana Montero (Zaragoza)

Sweden
D. Haffajee (Lund) I. Isaksson (Linkoping) E. Moberg (Göteborg)

Switzerland
A. Chamay (Geneva) A. Graedel (Schaffhausen) U. Heim (Coire) H. Ch. Meuli (Bern) V. Meyer (Zurich) H. Nigst (Basel) C. Simonetta (Lausanne) A. O. Narakas (Lausanne) I. Poulenas (Lausanne) C. Verdan (Lausanne)

Algeria
Y. Martini Benkeddache (Bainen Bologhine)

Iran
Goucheh (Teheran)

Israel
M. Rousso (Jerusalem)

Japan
M. Yashimura (Kanazawa City)

Libya
El Bacha

Argentina
E. Zancolli (Buenos Aires)

Venezuela
R. Contreras (Caras) E. Kamel (Caracas)

1. HISTOPATHOLOGY OF THE NAIL

G. Achten

The nail is a highly specialised keratinised appendage of the skin. It is also important in daily life:

it protects the ends of the fingers, which are constantly exposed to injuries

it contributes to the sensitivity of the ends of the fingers, a sensory territory of great importance in everyday activities

it has an aggressive role which may occasionally be called upon

it has an aesthetic function of some importance.

The nail grows on average between 3 and 4 mm per month. The nails on the hand grow more quickly than those on the feet. Growth is more rapid in the young, in the middle finger and in the dominant hand; it is least in the little finger and thumb. Seasons, professions, race and gender are not known to influence growth to any extent. General health, on the other hand, does; various endocrine disorders, avitaminosis, anaemia, physical fatigue, all slow down the rate of growth, as do certain toxic drugs.

The terminology and structure of different parts of the healthy nail and neighbouring tissues will first be described in order to establish the limits of normality. Against this background the pathology of the most frequently observed nail lesions will be described. A more detailed study of the nail's anatomy and pathology can be consulted elsewhere (Achten and Wanet, 1978).

TERMINOLOGY

The longitudinal cross-section (Fig. 1.1) shows the various components of the fingertip.

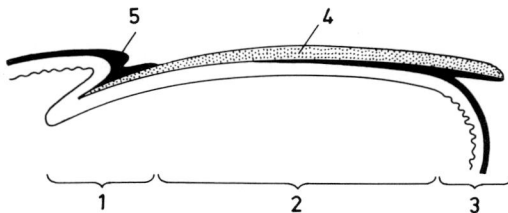

Figure 1.1
Longitudinal section of the finger end.
1. Nail matrix. 2. Nail bed. 3. Hyponychium. 4. Nail plate. 5. Eponychium.

MATRIX AT THE NAIL PLATE

The nail matrix, which started originally from an invagination of the epidermis of the back of the finger, extends from the end of this invagination to the distal extent of the lunula. It gives rise to the nail.

The nail proper consists of a hard, rectangular-shaped plate, convex dorsally. In the hands, the nail tends to be longer than it is broad and the opposite is true in the feet. The outer surface of the nail is smooth, whereas the under surface has longitudinal, parallel ridges, which match similar depressions on the nail bed. When the nail is avulsed, these striations are clearly visible.

NAIL BED

During growth, the nail slides on to the nail bed and becomes adherent to it. This area, derived from the dorsal epidermis, consists of a few layers of squamous cells. These adhere so tightly to the nail plate that, when the nail is avulsed, this layer remains attached to the under surface, thus exposing the underlying epidermis. While healing is taking place, the epidermis produces a hypertrophied corneal layer, which forms a 'false nail' on the nail bed. This keratinised formation is sometimes so well developed that it slows down or damages the growth of the new nail. It is then necessary to reduce this keratinised layer to ensure normal regrowth of the nail proper.

HYPONYCHIUM

The hyponychium extends from the distal end of the nail bed to the distal skin crease on the palmar aspect of the finger. It corresponds to the pulp of the finger.

EPONYCHIUM

The eponychium consists of the epidermal fold, which overlies the proximal part of the nail for a short distance forming the cuticle.

NAIL GROOVES

The nail grooves, which mark the limits of the nail, are shown in Figure 1.2. They may be separated into proximal, distal and lateral grooves, and are more clearly seen when the nail is avulsed. The distal groove situated on the hyponychium is covered by the nail plate, whereas the lateral and proximal grooves are clearly evident as they rise above the level of the nail. The nail is held in these grooves in the manner of a watch-glass.

LUNULA

The proximal part of the nail is made up of the lunula, a white area, whose convex end merges with the pink part of the nail. At

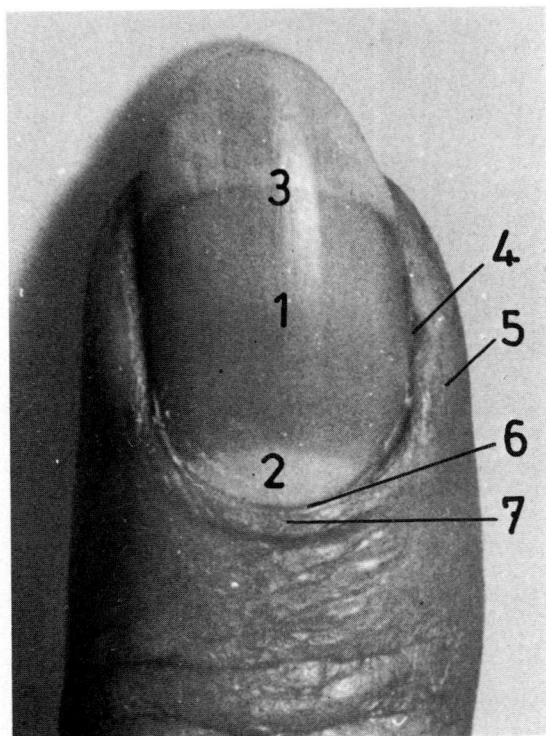

Figure 1.2
The normal nail.
1. Nail plate or nail proper. 2. Lunula. 3. Onychodermal junction. 4. Lateral groove. 5. Lateral fold. 6. Proximal groove. 7. Eponychium and proximal fold.

the free edge of the nail, there is a thin, pale, translucent line known as Terry's onychodermal band (Terry, 1955).

The nature of the white colour of the lunula is uncertain. From the views of various authors (Burrows, 1919; Ham and Leeson, 1961; Achten, 1963; Lewin, 1965; Baran and Gioanni, 1969), it seems that it is due both to the structure of the nail plate itself and to the specialised character of the underlying matrix.

At the lunula the nail is thin and flat and the epidermis is thicker so that the underlying capillaries cannot be seen through it. The nail is less firmly attached at the lunula than it is to the rest of the nail bed so that the light is reflected from the interface between the nail and this part of its bed.

TECHNIQUES

BIOPSY AND STAINING (FIG. 1.3)

1. *Longitudinal biopsy of the nail* (Zaias, 1967); a longitudinal fragment of the nail is removed under local anaesthetic (Fig. 1.3a). Using a scalpel, two parallel cuts are made from the eponychium to the distal end of the finger; the fragment is then detached with iridectomy scissors. This type of biopsy causes inevitable scarring, which may be unacceptable. The cut should

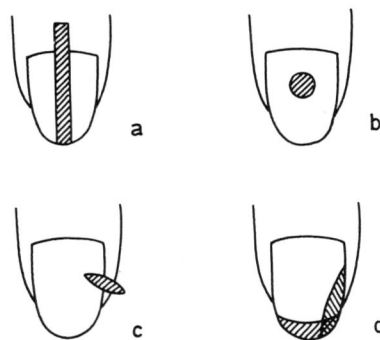

Figure 1.3
Different techniques of nail biopsy for histological studies.

never exceed 3 mm in width for this reason. It should not be used in cases of diabetes, scleroderma or vascular disease. Suture of the matrix at the nail bed reduces disfigurement. The advantage of longitudinal biopsy is that it allows examination of the matrix, nail bed and periungual tissues, as well as of the nail.

2. *In the 'punch biopsy'* small, round nail fragments are removed under local anaesthetic (Fig. 1.3b). The advantage of this technique is that it permits selective examination of any part of the nail. It has the disadvantage that it yields only a small amount of tissue for study and it is difficult to maintain orientation of the specimen during processing.

3. *Biopsy of the periungual fold* advocated by Stone and Mullins (1962) permits study of the pathology of the soft tissues, which are often affected, at the same time as the nail plate (Fig. 1.3c).

4. *Complete excision of the nail* is only justified when required for therapeutic reasons, as in cases of injury or ingrowing nail. Surgical amputations and autopsies allow a study of the nail *in toto*.

Pardo-Castello (1960) emphasises the difficulty of studying the pathology of the nail and the periungual tissues, because of the reluctance of most patients to permit an exploratory biopsy. It becomes necessary, therefore, to limit the study in these cases to fragments of the nail alone.

5. *The free end of the nail is removed* with nail clippers (Fig. 1.3d). According to the location of the lesion, the biopsy may include the complete width of the tip of the nail, or part of the margin. If possible, part of the hyponychium should be included with the fragment. The fragment is then imbedded in paraffin; fixation and dehydration are not required beforehand. A microtome is then used to cut the fragments into sections of 15 to 20 microns. The paraffin, which is used only as a support, strips off easily. The sections are then stained using either MacManus technique or toluidine blue. If necessary, other stains, such as haematoxylin, eosin, sulphydryl groups, ferric or melanic pigments, etc. can be used. The stained sections are then dehydrated in alcohol and cleared with toluol before mounting, as for normal, histological preparation (Achten, 1963) (Fig. 1.4).

This simple technique can only give limited information about the nail pathology since the biopsy does not include the matrix or nail bed.

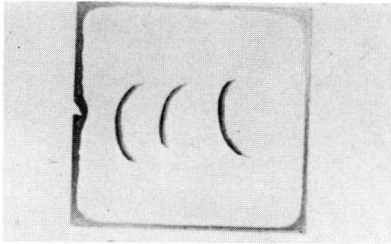

Figure 1.4
Histological cross-section of a nail fragment removed as in a manicure, embedded in paraffin without use of fixatives, cut into strips of 15–20 microns, which are stained and mounted as for a normal histological specimen.

VALUE AND LIMITATIONS OF HISTOLOGICAL DIAGNOSIS OF NAIL DISEASE

The nail to most clinicians is a *terra incognita* in relation to disease. Laboratory investigations, including histopathology, are of value but, if the examination is confined to a fragment from the tip of the nail, it may only reflect a problem which started months before in the matrix – that being the time necessary for the disease to spread from the base to the tip. The situation is further complicated by the fact that different nail diseases may have an identical effect on the keratin, whereas a single disease may sometimes produce variable changes in the different parts of the nail. From this it is apparent that nail histopathology cannot be interpreted in isolation from the clinical details.

THE HEALTHY NAIL

The fragment of healthy nail, when processed as above and stained with PAS, toluidine blue and the sulphydryl groups (Achten, 1963) reveals three layers with differential staining (Fig. 1.5).

THREE LAYERS OF THE NAIL TIP

The embryological, histological and histochemical observations of Lewis (1954), Alkiewicz (1964), Achten (1959, 1963, 1968), Zaias (1963), Jarrett and Spearman (1966), the electron microscopic studies of Hashimoto *et al*. (1966) as well as the experimental observations carried out on monkeys' nails following the injection of tritiated glycine (Zaias and Alvarez, 1968), all shed light on the nature of the different layers of the nail.

During the third foetal month an epidermal invagination of the dorsal end of the fingers forms the preliminary stage of the nail structure. The inferior surface of the epidermal fold forms the matrix or germinal layer (Fig. 1.6). The most proximal cells form the superficial layer of the nail, while the distal cells form the deeper layer. As the nail grows distally and comes to rest on the nail bed, a layer of keratin from the nail bed attaches itself to the under surface of the nail. This keratin, which is not an integral

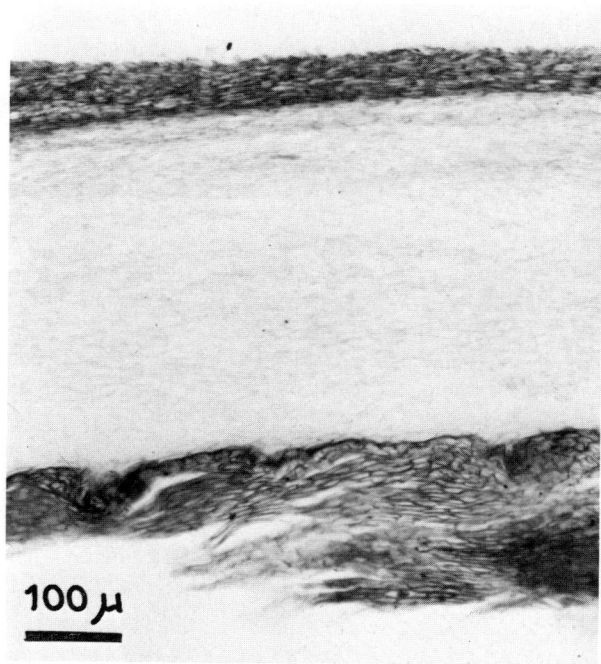

Figure 1.5
The healthy nail and its three layers. The superficial layer stained by MacManus technique; the deeper layer, which does not accept stain, and the hyponychial keratin, which becomes strongly stained and adheres tightly to the under-surface of the nail.

part of the nail, migrates peripherally with it and remains firmly attached to it. The three layers are illustrated diagrammatically in Figure 1.6.

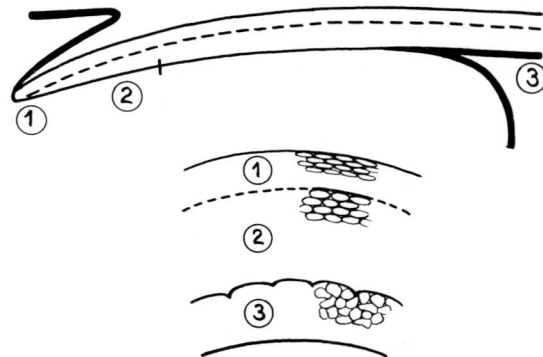

Figure 1.6
Diagram showing the origin of the three nail layers.
1. The proximal part of the matrix forms the superficial layer.
2. The distal part of the matrix forms the deeper layer.
3. The hyponychial keratin forms the keratinised area, which is attached to the under-surface of the nail. The distribution of cells and their staining characteristics differ in each layer.

Several authors (Lewis, 1954; Achten, 1963; Samman, 1961) have called these three areas the dorsal, intermediate and ventral layers, which could give the impression that the nail itself is composed of three parts; it is more accurate, however, to refer to superficial and deep layers of the nail and to the keratin of the nail bed as the hyponychium. The hyponychial keratin may become hypertrophied in certain disorders.

The cells of these three nail zones have a different physico-chemical nature and different staining properties. The cells of the nail proper are seen both in transverse and longitudinal sections to be arranged regularly, interlocking like tiles on a roof. Their main axis is horizontal. The cells of the superficial layer are closer together and flatter than those of the deeper layer. In the hyponychial keratin, the corneal cells are more polyhedral, like the structure of the malpighian cell, and are less regularly arranged.

Keratin is a scleroprotein formed by regularly repeated groupings of amino acids, which form the main longitudinal chains of the keratins. These are joined to each other by lateral bonds (disulphide; acid-base; hydrogen; and Van der Waals linkages). There are, however, several types of keratins ranging from the soft keratin of the skin epidermis to the hard keratin of nails and hair. These differ from each other in the composition of the main chains and in the number of lateral bonds; the latter being more numerous in the hard keratins (Fig. 1.7). If the

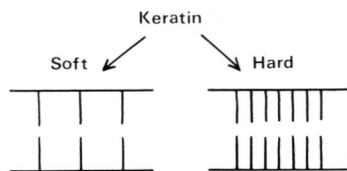

Figure 1.7
Structure of the keratins. The lateral bonds, which join the main chains are much fewer in the soft keratin (skin epidermis) than in the hard keratin (hair and nails).

lateral bonds are numerous, then the free radicles are few and cannot in consequence combine readily with different stains. Softer keratins, therefore, are more readily stained than the harder forms and this explains the differential staining characteristics of the superficial layers of the nail (Fig. 1.8):

   (a) the superficial layer made up of moderately hard keratin
   (b) the deeper layer made of hard keratin
   (c) the hyponychial layer made of soft keratin like that of the skin.

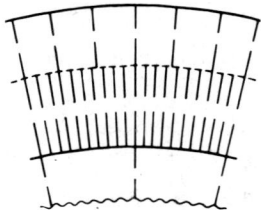

Figure 1.8
The three layers of the nail and their physico-chemical constitution showing the number of lateral bonds. The superficial layer is made of moderately hard keratin. The deeper layer is made of very hard keratin. The hyponychial layer consists of soft keratin.

INTERPRETATION OF HISTOLOGICAL LESIONS

An interpretation of a transverse cross-section of a nail must take account of the clinical problem. It is necessary to know whether the lesion started proximally or in the lateral or distal part of the nail, how it has developed and what treatment has been given.

Figure 1.9 shows that a lesion in the eponychial or matrix area will affect first the root of the nail and spread gradually to the nail end. Whereas, if the lesion is situated in the nail bed, it will be visible through the nail and may eventually damage it; it will start in the middle of the nail and migrate towards the end of the finger as the nail grows. If the hyponychial area is affected, the keratin will be attacked at the distal end but the disease may extend proximally to involve the matrix.

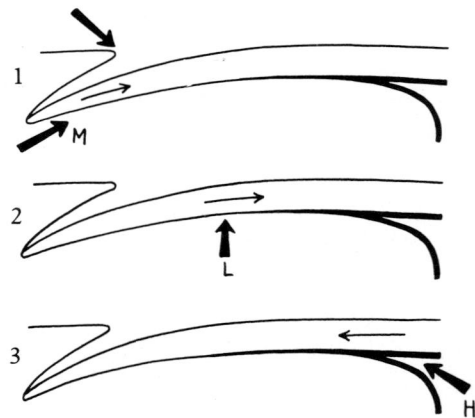

Figure 1.9
The clinical evolution of nail lesions according to site of origin.
1. Involvement of the matrix or the eponychial area will damage the keratin, originating in the nail matrix, and will spread with the growth of the nail to the distal end.
2. Involvement of nail bed produces a lesion within the nail plate, which spreads to the distal end.
3. Involvement of hyponychium produces a lesion, which starts at the free edge of the nail and may extend proximally.

The changes observed in a transverse cross-section of the three nail layers indicate where the original lesion began (Fig. 1.6). Damage to the superficial layer means either that there is a disorder of the proximal part of the matrix, or else that this layer was exposed to chemical action or surface infection. If only the deeper layer is involved, this generally indicates damage to the distal part of the matrix. Changes seen in the hyponychial area mean either that the nail bed is diseased or that it is exposed to chemical or infective injury.

Involvement of more than one layer may occur, for example, in dermatological diseases which effect the matrix of the nail bed. Dermatophytes and chemical irritants usually attack the most vulnerable keratin, that is of the hyponychial area, then the superficial layer and finally the deeper layer.

These observations, coupled with the clinical history, assist the pathologist to identify the original lesion.

THE PATHOLOGICAL NAIL

This study of the pathological nail will be restricted to the most common diseases. Nail diseases can be classed into two groups – congenital and acquired diseases. Acquired diseases are divided into five groups.

1. Dermatological disorders which involve the nail, such as psoriasis, eczema, erythroderma, lichen planus, alopecia areata and radiodermatitis; of these psoriasis will be described.
2. Infections such as onychomycoses due to *Candida*, and dermatophytes, paronychia and onychia, due to bacterial infections; the onychomycoses will be described.
3. The onychodystrophies which accompany general disease and those of undetermined origin; the most common being:
   (a) finger clubbing or the watch-glass nail
   (b) onycholysis with pachyonychia, characterised by a hypertrophy of the keratin of the hyponychium leading to progressive thickening and consequent elevation of the nail from the bed; onycholysis will be described
   (c) onychorrhexis characterised by the presence of longitudinal parallel grooves on the dorsal surface of the nail.
4. Nail discolouration, which include leukonychia, haematoma and nail melanoses; leukonychia will be described.
5. Nail tumours of which melanoma and malignant melanoma are the most important.

CONGENITAL DISEASES

These disorders may be confined to the nail or may be associated with disorders of the skin or other tissues. The diseases in this group are diverse and rare, among them are: anonychia, hyponychia, pachyonychia, leukonychia, leukokoilonychia, ichthyosis and bullous epidermolysis. The effect on the nail is often gross involving the nail plate, the hyponychial keratin, or both. An association of leukonychia with koilonychia (concave or spoon nail), observed by Baran (1969)

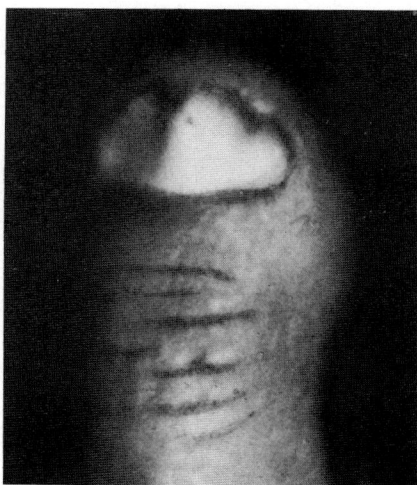

Figure 1.10
Congenital leukokoilonychia (case of Dr. Daran). The nail is both white and concave (like a spoon).

illustrates this group of disorders. The mother has associated leukokoilonychia of the feet and hands; the son has koilonychia of the index finger, thumbs (Fig. 1.10), and the big toes.

Microscopically, the nail structure is completely altered (Fig. 1.11). The cells have lost their usual polarity; the end of the nail

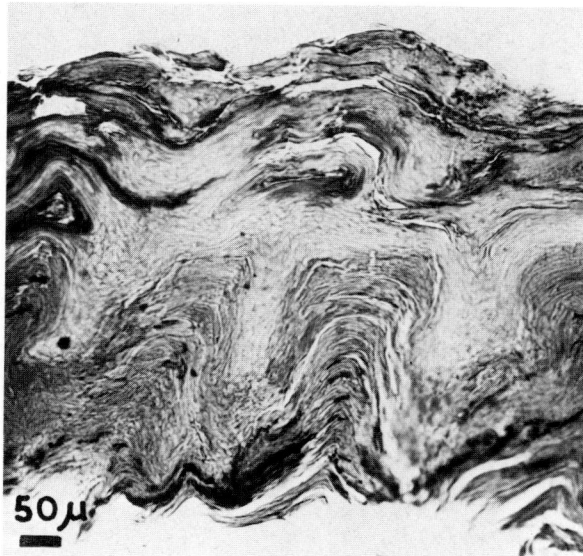

Figure 1.11
Congenital leukokoilonychia. The three nail layers appear irregular. Undulation and interlocking occur most frequently between the hyponychial keratin and the deep layer (MacManus).

is furrowed as much in the superficial surface as on its hyponychial aspect. The hyponychial area is thick; note the interlocking of the hyponychial and nail plate keratins. Foci of parakeratosis extend widely onto the nail itself. These structural changes are best seen by polarised light. The regular disposition of the fibrils responsible for the well-ordered birefringence of the healthy nail (Fig. 1.12) gives way to alternating birefringent and

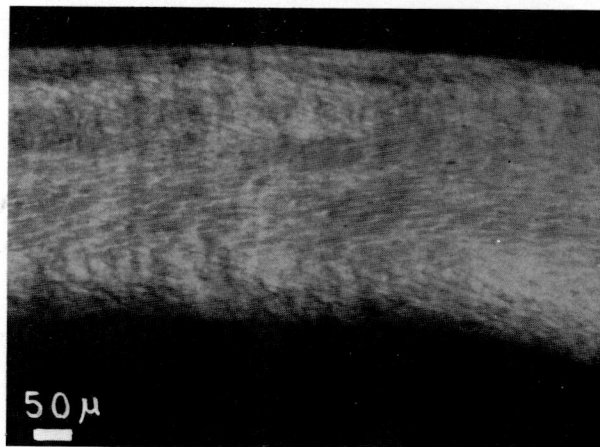

Figure 1.12
Healthy nail. Regular appearance of the birefringence of the keratin fibres seen under a polarising microscope.

dark bands indicating disorientation of the keratin fibres (Fig. 1.13). In this case, the fault lies in the matrix, nail bed and the hyponychium.

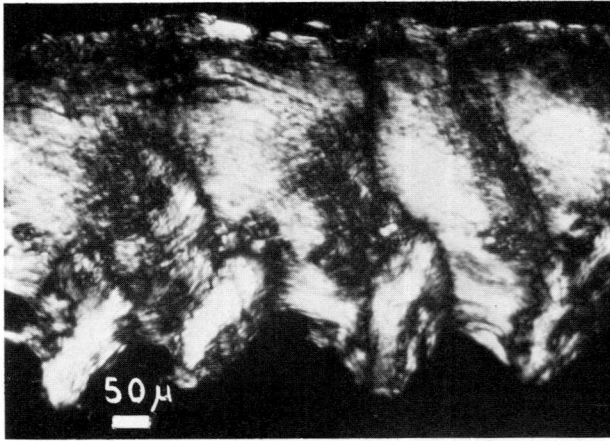

Figure 1.13
Congenital leukokoilonychia. Irregular birefringence of the fibres of the keratin show a change in their orientation.

ACQUIRED DISEASES

Nail Dermatoses. Many dermatoses produce ungual lesions, including eczema, psoriasis, lichen planus, alopecia areata, radiodermatitis and various bullous disorders. In this chapter psoriasis alone will be studied.

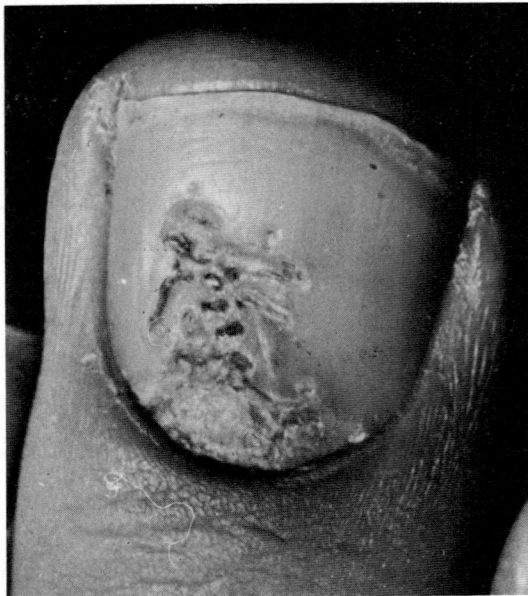

Figure 1.14
Psoriasis of the nail. Usual clinical appearance: the lesion starts in the proximal area and spreads to the end of the nail. Small pits give the nail the appearance of the so-called 'thimble nail'.

Nail psoriasis. The most frequent lesion in psoriasis is hypertrophy of the nail bed, which raises up the nail plate; the nail separates from the hyponychium at its free edge resembling a mycotic infection. Also common are punctate erosions, which form the 'thimble nail' (Fig. 1.14). Neither of these lesions is pathognomonic for they can occur in other skin diseases. Histology of the psoriatic nail may show:

   (a) hypertrophy of the nail bed
   (b) parakeratosis of the nail plate, involving the superficial and deep layers

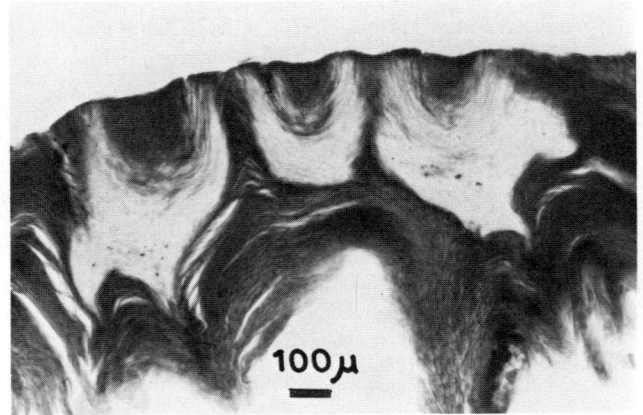

Figure 1.15
Psoriasis of the nail. Psoriatic pits excavating the surface of the nail (MacManus).

   (c) pitting on the nail surface (Fig. 1.15); these pits being filled with a keratin that is strongly stained by the MacManus technique and by toluidine blue. Frequently there are densely stained concentric bands under these pits, the cells of which are parakeratotic (Fig. 1.16); the soft keratin of these pits can disappear, leaving in its place excavations, which correspond clinically to Milian's punctate erosions

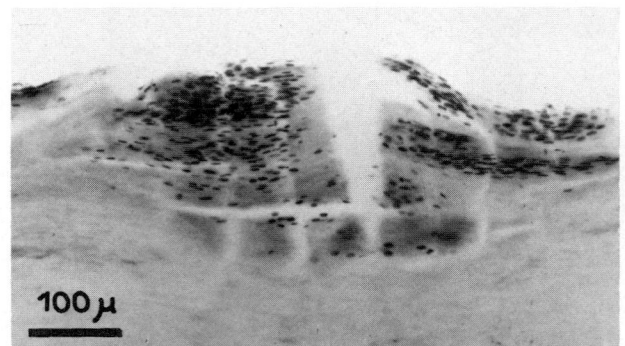

Figure 1.16
Psoriasis of the nail. The cells which produce the 'thimble' pits keep their nuclei (parakeratosis) separate from one another and finally become detached from the nail.

   (d) in the most advanced cases, the whole structure of the nail plate is completely altered (Fig. 1.17). Keratinised

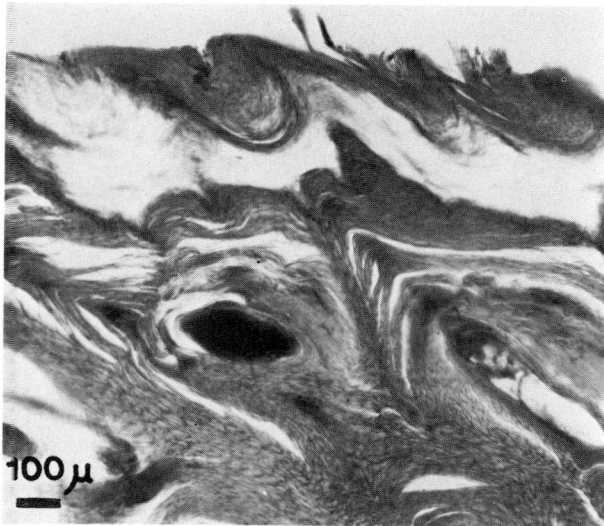

Figure. 1.17
Psoriasis. Extensive alteration of the nail architecture, with total disorganisation; the hyponychial keratin is hypertrophied.

processes, nearly always parakeratotic, form in the substance of the deeper layer, and foci of parakeratosis may occur throughout.

In order to understand these lesions in the psoriatic nail, it is helpful to recall the process of accelerated skin growth of cutaneous psoriasis (Fig. 1.18a). In the healthy epidermis, the cell starting from the basal layer takes on average, according to its site, twenty to twenty-seven days to reach the surface and develop into a squamous cell. In psoriasis this same cell takes three to four days to reach the surface (Rothberg *et al.*, 1961).

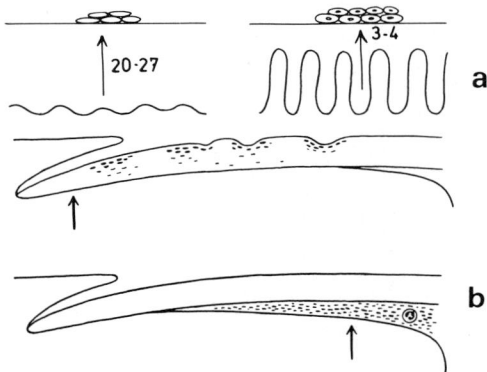

Figure 1.18
(a) Histopathogenesis of psoriatic lesions. The basal cell of the epidermis takes on average 20 to 27 days to become a squamous cell. In psoriasis, this epidermopoiesis is speeded up, the time being reduced to 3 to 4 days. The result is the formation of immature squamous cells, which retain their nuclei. Similar involvement of the matrix in the psoriasic nail leads to pockets of immature parakeratotic cells, which separate producing the psoriasic pits.
(b) Involvement of the nail bed produces a hypertrophy of the hyponychial keratin, which becomes parakeratotic and elevates the nail plate. Munro's abscesses, sterile abscesses consisting of focal accumulation of leucocytes, also occur.

This speed of progression means that the epidermal cell cannot become mature and form completely structured keratin. This is shown by the persistence of the nucleus, which is characteristic of the parakeratosis of psoriasis. This also explains the presence of certain enzymes which are not normally found on the corneal layer (Braun-Falco, Thianprasit and Kint, 1963). The changes in the psoriatic nail can also be attributed to over-rapid maturation of the epidermal cells with persistence of nuclei, and hence parakeratosis, which occurs in the superficial and deep layers and is accompanied by the formation of immature keratin which remains soft. These lesions may occur in isolated foci and are the source of nail pits, since this softer keratin separates easily from the rest of the nail.

Onycholytic lesions, in which the nail bed gets thicker and the nail becomes detached, are attributable to the accumulation of parakeratotic layers formed by the epidermis of the nail bed as in the cutaneous lesions of psoriasis (Fig. 1.18b).

Pustular psoriasis is characterised by the presence in the parakeratotic layer of numerous leucocytes, which form sterile micro-abscesses, and this is also found in the nail bed (Figs. 1.19 and 1.20). This may occur also in other diseases in which similar sterile pustules described by Kogoj are found.

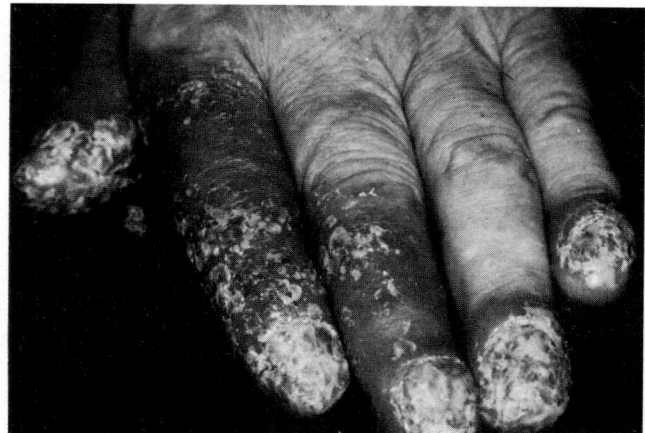

Figure 1.19
Pustular psoriasis. Munro's abscesses enlarge to produce sterile pustules.

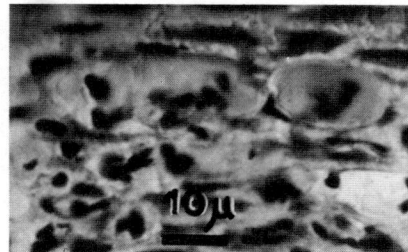

Figure 1.20
Psoriasis. Munro's abscess consisting of leucocytes and squamous cell-debris.

NAIL INFECTIONS – ONYCHOMYCOSES

Among the nail infections, the most important are the onychomycoses caused by dermatophytes and *Candida*, and pyogenic infections of the perinychium and nail bed. The onychomycoses (Fig. 1.21) will be described. Onychomycosis

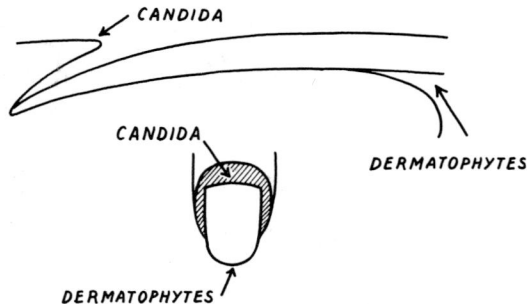

Figure 1.21
Diagram of nail diseases of fungal origin. Dermatophytes, which are keratinophilic, involve the nail keratin at the end of the nail, producing dermatophytic onychia. *Candida* causes paronychial lesions, involving the lateral and proximal nail grooves.

due to dermatophytes starts in the end or the side of the nail and progresses from there towards the matrix. Onychomycosis due to *Candida* affects the lateral and proximal nail grooves; the damage to the latter involves the matrix and spreads to the end of the finger as the nail grows.

Dermatophytic onychomycosis. This can attack one or more nails and usually starts at the free edge (Fig. 1.22) or lateral borders of

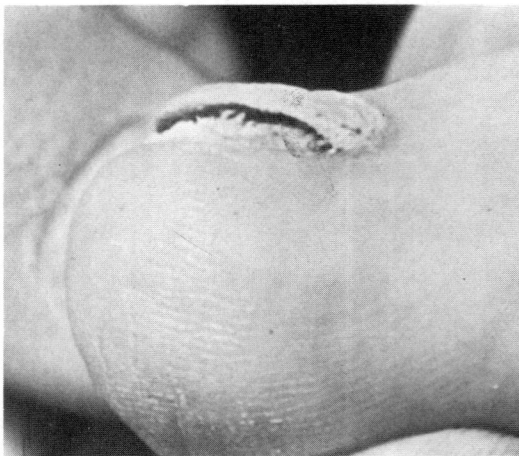

Figure 1.22
Initial stages of dermatophytic onychomycoses. The nail is slightly elevated from its bed (onycholysis). This elevation is caused by the hyponychial keratin, which, when infected, becomes hypertropnied and friable.

the nail, producing a thickening of the bed in the form of a corneal mass, which pushes up the nail and detaches it from its bed in its distal part. The detached part becomes yellowish or greyish-white in colour. A lesion starting on the lateral border of

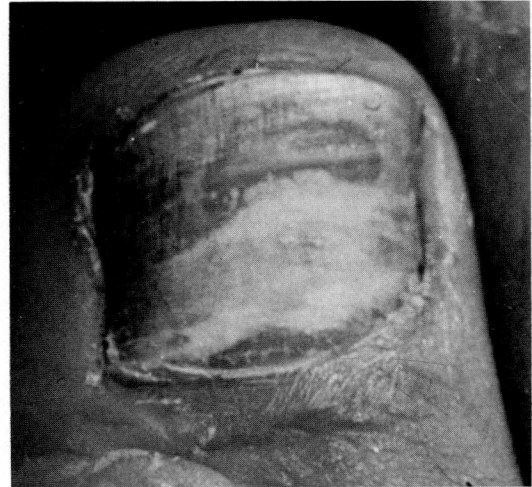

Figure 1.23
Dermatophytic onychomycosis forming a transverse band starting from one lateral groove and extending to the other.

the nail can form transverse extensions, which spread gradually towards the opposite border (Fig. 1.23). The nail becomes gradually dull, rough, friable and eroded (Fig. 1.24).

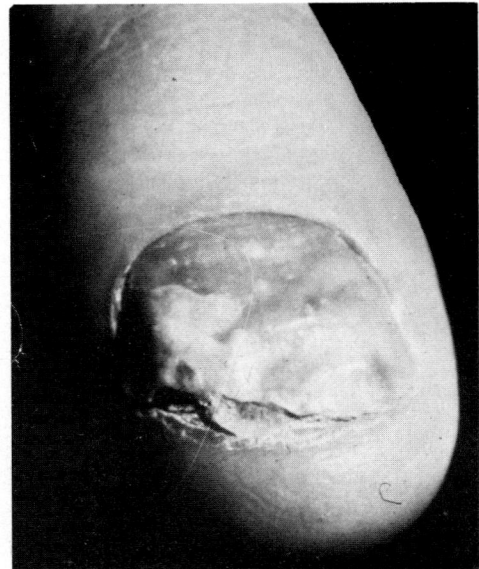

Figure 1.24
Dermatophytic onychomycosis with progressive destruction of the nail.

Histologically, the MacManus stain seems best for the detection of ungual mycosis. Several authors, including Kligman, Mescon and Delamater (1951), Gadrat, Bazex and Dupré (1952), Jillson and Piper (1957), Prunieras and De Beer (1960, 1961), have pointed out the effectiveness of histochemical stains for the detection of mycelia filaments. In the majority of cases of onychomycosis due to dermatophytes, the hyponychial

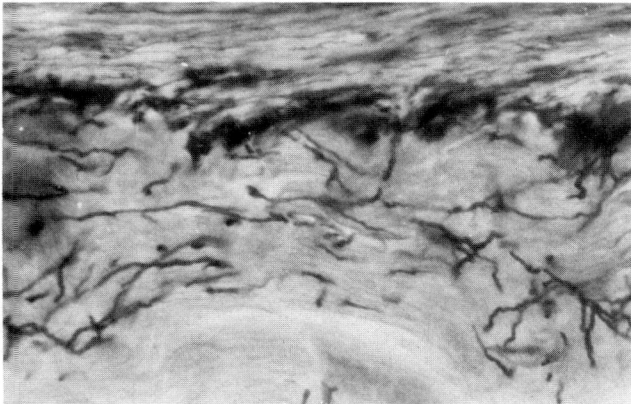

Figure 1.25
Dermatophytic onychomycosis. Mycelium filaments spread in all directions in the hyponychial keratin.

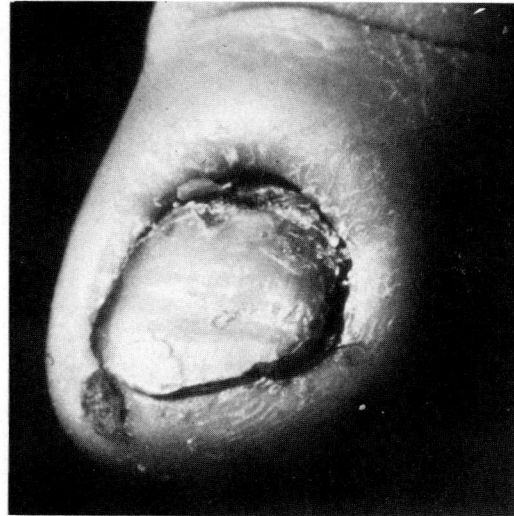

Figure 1.27
Paronychia due to *Candida albicans*, oedema of the lateral folds, secondary damage to the nail. The brownish colour of the folds is a characteristic but inconstant feature.

keratin is invaded either by a mat of filaments or by widely-radiating filaments (Achten and Simonart, 1963) (Fig. 1.25). For cultural studies keratin derived from the nail bed is best. Filaments can also be seen in the nail itself, particularly in onychomycoses which have been developing for several months or years. These filaments are regularly arranged parallel to the transverse axis of the nail (Fig. 1.26).

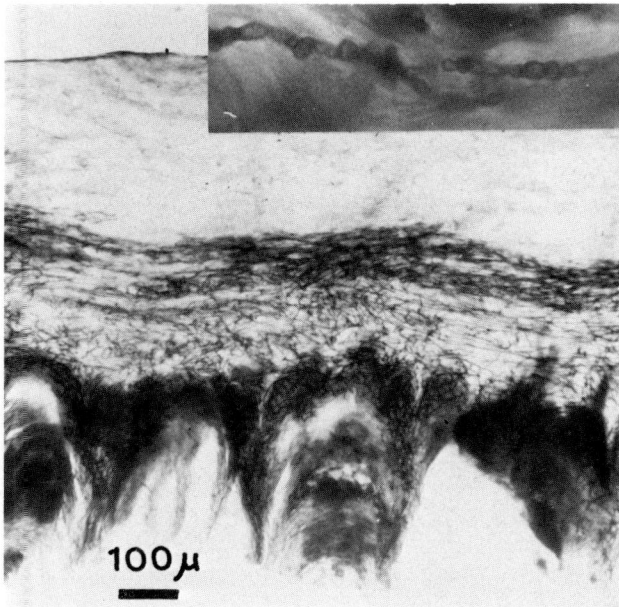

Figure 1.26
Dermatophytic onychomycosis. Mycelium filaments run parallel to the surface of the nail within the nail plate itself. In the inset, note the segmented appearance of the filaments.

Onychomycosis due to Candida. This appears most often as a chronic paronychia (Fig. 1.27) and may affect several digits, most commonly the fingers. The lateral borders or eponychium

are surrounded by a red oedematous groove, which is painless or only slightly painful. The nail itself may be secondarily involved and candidiasis of the nail without paronychial lesions has been described. In such cases laboratory tests are necessary to distinguish onychomycosis due to dermatophytes from that due to *Candida*.

The localisation of the infection determines the histological picture: paronychia of the proximal groove can produce structural modifications secondary to the inflammation in the nail but yeasts can only rarely be identified. On the other hand, when the lateral furrows are attacked, it is possible to remove the lateral part of the nail in which yeasts can be seen in the hyponychial keratin (Fig. 1.28). They may be linked by pseudo-filaments, which can be observed in both the hyponychium and the nail plate (Fig. 1.29).

Studies of the invasion of the ungual keratin by dermatophytes

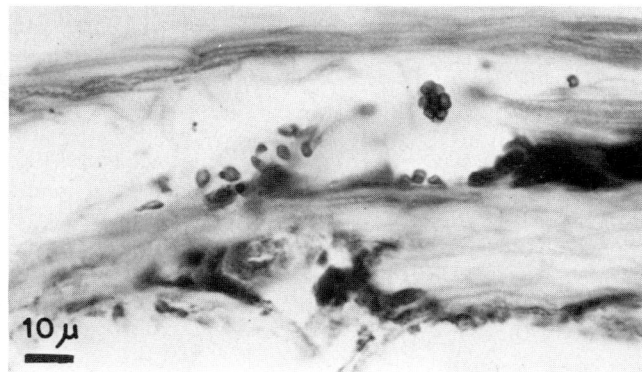

Figure 1.28
Candida onychomycosis. *Candida* in its yeast form is separating the squamous cells of the nail.

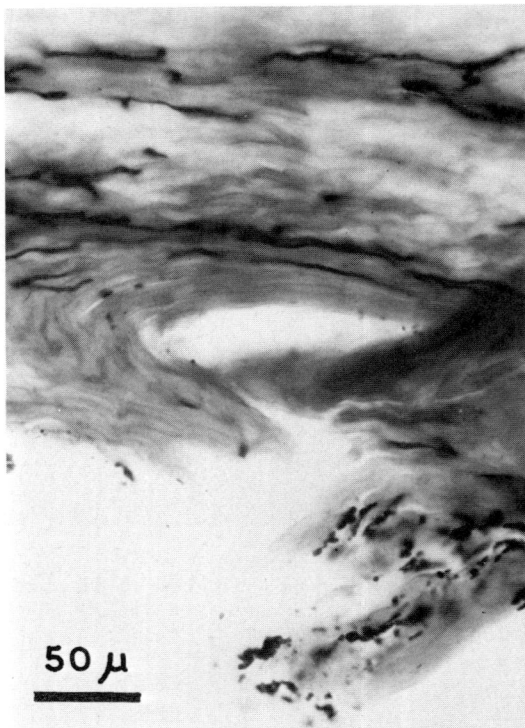

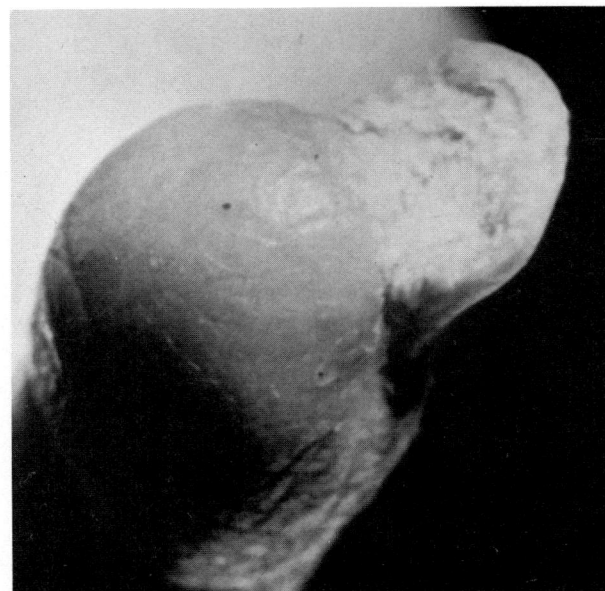

Figure 1.30
Onycholysis. Thickening of the hyponychial keratin, which is pushing up the nail. The latter looks thickened (pachyonychia).

Figure 1.29
Candida onychomycosis. *Candida* in its yeast form inferiorly and in its pseudo-filamentous form in the nail.

and yeasts have been carried out *in vitro*. The hyponychial keratin is first involved as the soft keratin is more easily attacked than the hard keratin of the nail, both by dermatophytes and by yeasts of the *Candida* group (Achten and Simonart, 1964, 1965; Roobaert and Achten, 1969).

ONYCHODYSTROPHIES

The onychodystrophies are a diverse group of disorders which may be associated with general or cutaneous disorders. Somtimes their origin is obscure.

Onycholysis. In onycholysis with thickening of the nail, or pachyonychia, involving the keratin of the hyponychium and the nail bed (Fig. 1.30), rounded or oval amorphous masses can be seen in these areas. Often these are small, but the confluence of several small masses may produce a larger one. These homogeneous masses are often piled on top of each other (Fig. 1.31). They are surrounded by healthy corneum and are usually separated from each other by clear spaces, giving a honeycomb appearance in severe cases (Fig. 1.32). The contents of these intercellular spaces have staining characteristics, which suggest that they are formed of mucopolysaccharides (Achten and Wanet, 1970).

NAIL DISCOLOURATION

Nail discolourations depend on many factors including

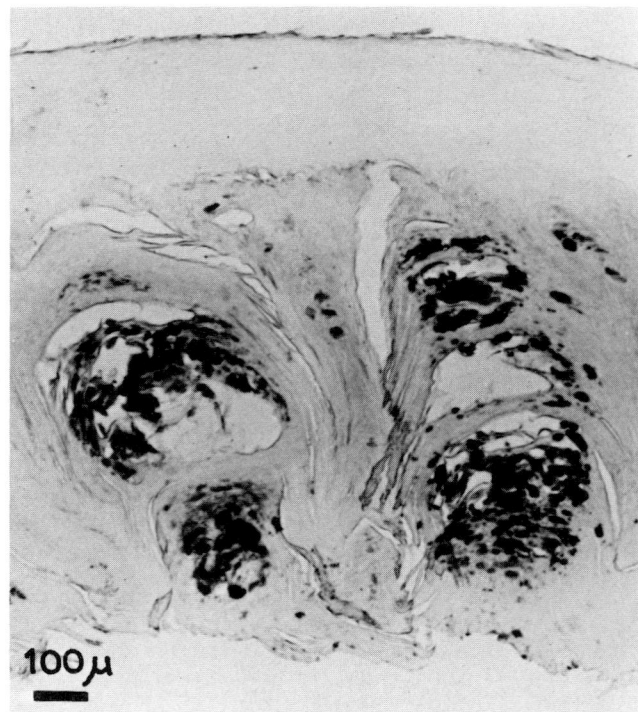

Figure 1.31
Onycholysis with pachyonychia. The nail plate is healthy. The hyponychial keratin is hypertrophied. The stained areas indicate the presence of mucopolysaccharides (Hale's stain).

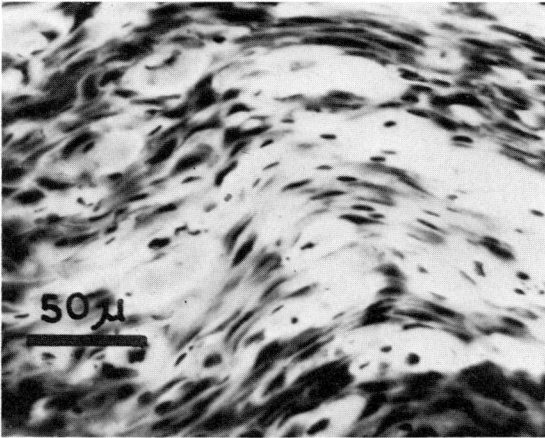

Figure 1.32
Onycholysis with pachyonychia. At a later stage of the disease loss of pockets of mucopolysaccharides leaves a honeycomb appearance.

modifications to the keratin of the nail plate, damage to the underlying tissues, separation of the nail from its bed and deposits of various substances. The most common are leukonychia, haematomas and ungual melanoses.

Leukonychia. This usually appears as white spots or striae (Fig. 1.33). More rarely, it involves the whole nail (Fig. 1.34). It can be congenital or acquired; the latter having a range of pathological appearances.

There are several histological explanations of this disorder: (Heidingsfeld, 1900; Sibley, 1922; Eller and Anderson, 1928; Becker, 1930; Alkiewicz, 1935; Singer, 1937; Mitchell, 1953;

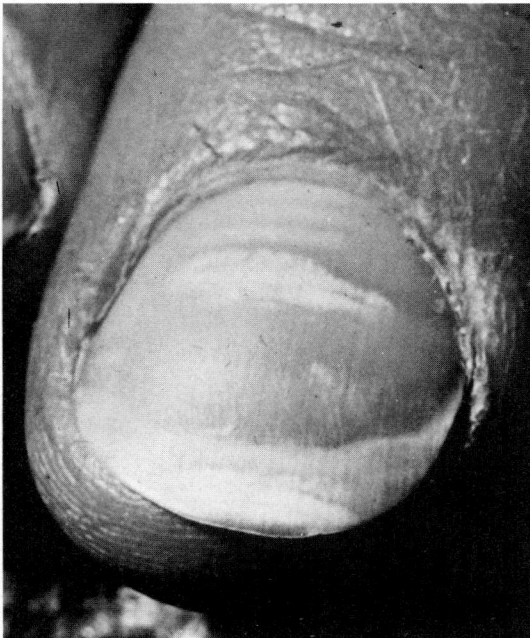

Figure 1.33
Striated leukonychia.

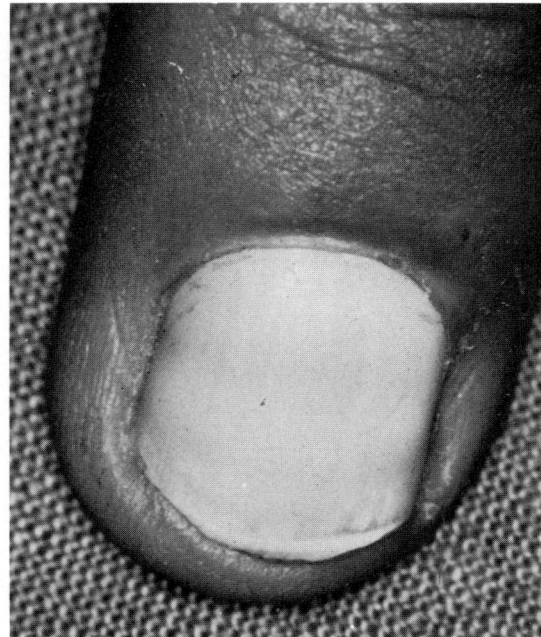

Figure 1.34
Total leukonychia.

Pardo-Castello, 1960; Lewin, 1965; Samman, 1965). The presence of cavities filled with air, abnormally nucleated cells, areas of hyalinised keratin, yellowish intracytoplasmic granules, and fissuring of the nail have all been described.

We have seen leukonychia associated with hyperchromatic bands in the nail plate. The corneal cells were more flattened than usual (Fig. 1.35). The nature of the yellowish-white granules, noted by Alkiewicz, have not been determined by histochemical stains (Fig. 1.36). We have not seen the formation of cavities or grains of hyalinised keratin. There are many grounds for thinking that leukonychia is related to a fault in keratinisation and the structural arrangement of the squamous cells.

UNGUAL TUMOURS

Amongst the nail tumours, melanoma and malignant melanoma, or 'melanotic whitlow', are important because of the therapeutic problems that they pose.

The matrix and nail bed contain melanocytes (Alkiewicz, 1958; Higashi, 1968; Higashi and Tade, 1969). In white people, the nail plate does not contain or only very rarely contains melanic pigment. In black people, a pigmentation in the form of striae is often seen and occasionally diffuse pigmentation occurs. After administration of MSH or after radiotherapy, non-functional melanocytes can become active (Pillsbury, Shelley and Kligman, 1956).

Ungual melanoma is always junctional; the pigmented cells are found in the junction of the epidermis and the dermis (Fig. 1.37). According to Duperrat and Cintract (1960), Duperrat and

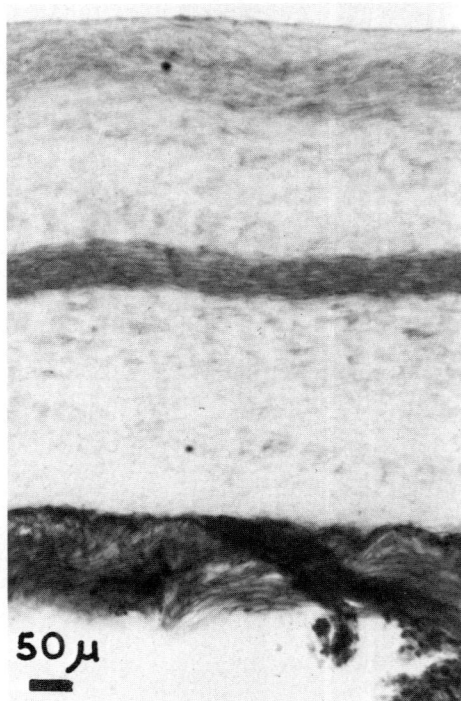

Figure 1.35
Striated leukonychia. The hyperchromatic band in the nail points to a change in the process of keratinisation (MacManus).

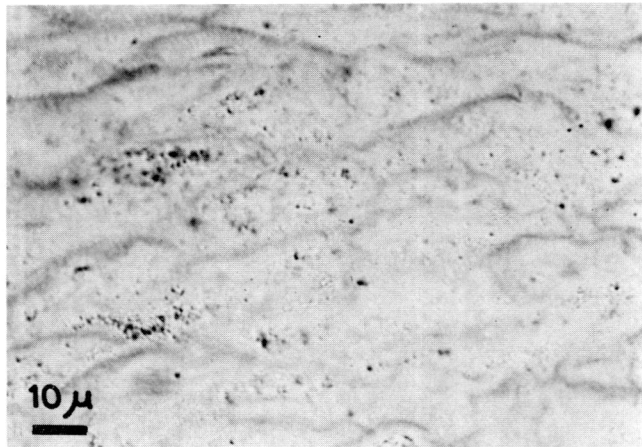

Figure 1.36
Leuconychia. Refringent brownish-yellow grains are present, the nature of which could not be established by staining techniques.

Mascaro (1966), and Pack and Oropeza (1967), this lesion is only a stage preceding malignant transformation induced by trauma. Duperrat and his colleagues have given an authoritative description of ungual melanomas. Melanoblasts containing pigment can be seen amongst the basal cells. Giant cell forms may also occur (Fig. 1.38). The presence of giant cells, which infiltrate deeply and destroy the surface epithelium, points to

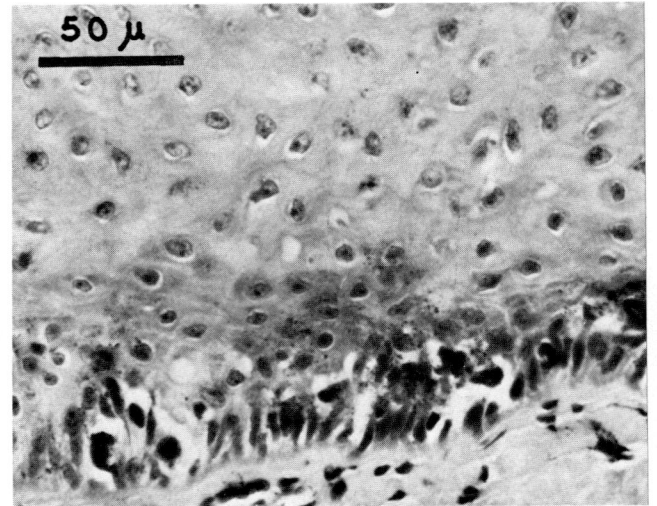

Figure 1.37
Ungual melanoma. Numerous pigmented cells are present on the matrix characterised by the clear zones surrounding them.

malignant change. A chronic inflammatory reaction is always present.

Between these extremes come all stages of transition, making histological assessment difficult. These lesions appear on the nail as a longitudinal, pigmented band (Fig. 1.39), as a more diffuse pigmentation, as nail deformities, as a subungual melanin stain, as a lateral detachment of the nail or as melanotic paronychia, described by Hutchinson in 1886 as melanotic whitlow (Fig. 1.40).

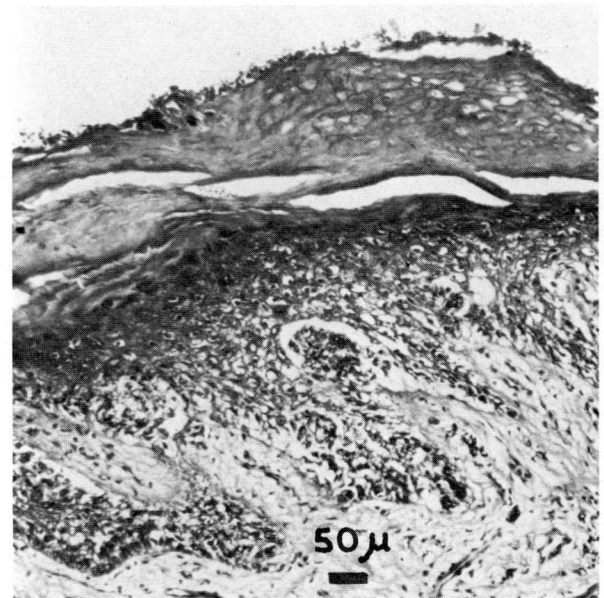

Figure 1.38
Ungual melanocarcinoma (junctional type). The tumour extends to the lower surface of the epidermis. The dermis and overlying epidermis, which are separated, have been invaded. Giant cell forms occur.

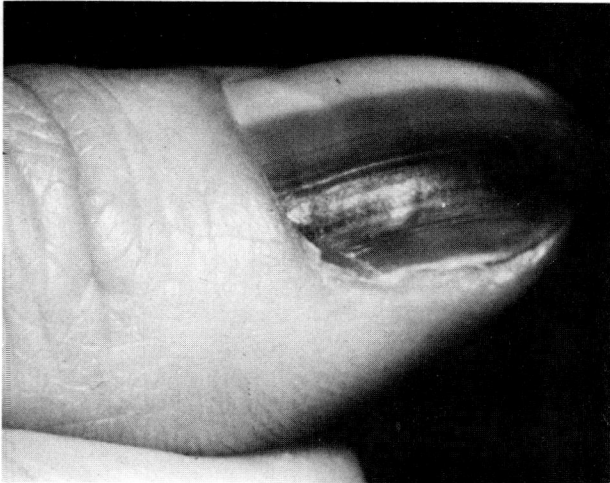

Figure 1.39
Ungual melanoma. A pigmented band occurs on the nail.

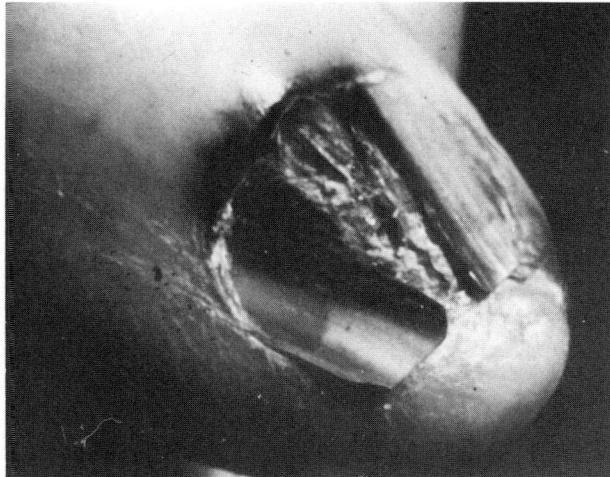

Figure 1.40
Ungual melanocarcinoma. The nail is deformed and damaged by the tumour. The involvement of the proximal fold (Hutchinson's melanotic whitlow), is very clear.

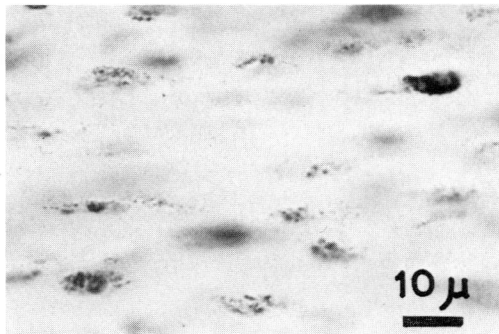

Figure 1.41
Melanin granules in the corneal cells in the nail.

Melanin pigment, which can be differentiated from iron pigment, can be seen clearly in the distal fragment of the nail plate (Fig. 1.41). Melanin pigment in the nail plate can be seen also in Addison's disease (Conard and Achten, 1974) (Fig. 1.42) and in radiodermatitis (Fig. 1.43); hence the importance of assessing the histology with knowledge of the clinical background.

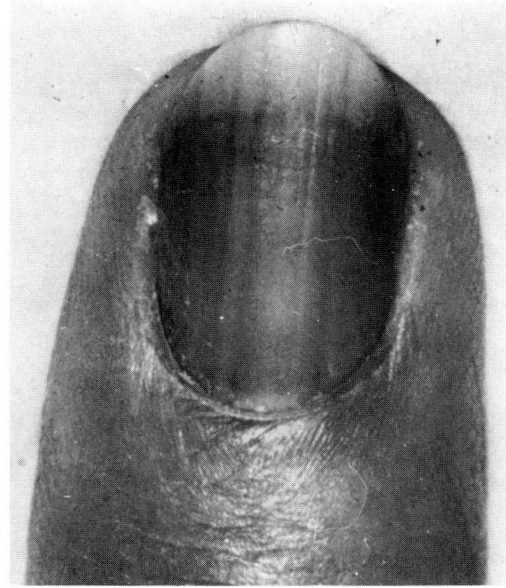

Figure 1.42
Addison's disease characterised by a diffuse melanotic pigmentation of the nail. The pigmentation may also occur as wide bands.

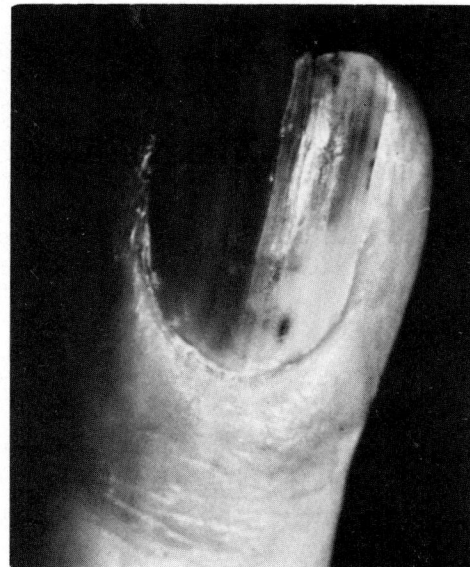

Figure 1.43
Radiodermatitis. Both ferric and melanotic pigment occur.

CONCLUSION

By systematic study of the healthy and the pathological nail the pathologist may assist in diagnosis. Too many patients whose histological picture is that of psoriasis have been subjected earlier to inappropriate treatment, when a diagnosis of ungual mycosis has been made without laboratory confirmation.

To illustrate the pathology of the nail, examples have been selected to show that nail keratin is the key to some of the mysteries. However, histology also has its limitations and the pathologist may not be able to resolve all the difficulties. Let us finish with these words of Foch: 'Do not tell me that this is a difficult problem. If it were not difficult, it would not be a problem.'

BIBLIOGRAPHY

[1] ACHTEN G. — Recherche sur la kératinisation de la cellule épidermique chez l'homme et le rat. *Arch. Biol.*, 1959, *70*, 1–119.

[2] ACHTEN G. — L'ongle normal et pathologique. *Dermatologica*, 1963, *126*, 229–245.

[3] ACHTEN G. — Normale histologie und histochemie des Nagels. Handbuch der Haut und Gesclechtskrankeiten. *J. Jadassohn*, 1968, *1*, 339–376.

[4] ACHTEN G., SIMONART J. — L'ongle: étude histochimique et mycologique. *Ann. Derm. Syph. (Paris)*, 1963, *90*, 569–586.

[5] ACHTEN G., SIMONART J. — Invasion des ongles par les dermatophytes in vitro. *Ann. Soc. Belge Méd. Trop.*, 1964, *44*, 755–766.

[6] ACHTEN G., SIMONART J. — Kératine unguéale et parasites fungiques. *Mycopathologia (Den Haage)*, 1965, *27*, 193–199.

[7] ACHTEN G., WANET J. — Pachyonychia. *Brit. J. Derm.*, 1970, *83*, 56–62.

[8] ACHTEN G., WANET J. — Pathologie der Nagel. *Spezielle Pathologische Anatomie Band 7 (2e éd.)*, 487–528, Heidelberg, Springer-Verlag, 1978.

[9] ALKIEWICZ J. — Klinik und Histopathologie der Leukonychie. *Przegl. derm.*, 1935, *30*, 1–36.

[10] ALKIEWICZ J. — Zür Histopathologie der Nagelmelanose (melanosis unguis). *Z. Haut u. Geschl., K.*, 1958, *24*, 14–17.

[11] ALKIEWICZ J. — Pathologischen Reaktionen an den epithelialen Anhangsgebilden: Nägel. Handbuch der Haut und Geschlechtskrankheiten. *J. Jadassohn*, 1964, *1*, 299–343.

[12] BARAN R., ACHTEN G. — Les associations congénitales de koïlonychie et de leuconychie totale. *Arch. belg. Derm. Syph.*, 1969, *25*, 13–29.

[13] BARAN R., GIOANNI T. — Les dyschromies unguéales. *Hôpital (Paris)*, 1969, *57*, 101–107.

[14] BECKER S. — Leuconychia striata: report of a congenital cases. *Arch. Derm. Syph. (Chic.)*, 1930, *21*, 957–960.

[15] BRAUN FALCO O., THIANPRASIT M., KINT A. — Uber den Einfluss einer lokalen Okklusiv-Therapie mit Fluorandrenolon auf die psoriatische Hautreaktion. *Arch. klin. exp. Derm.*, 1963, *217*, 30–49.

[16] BRAUN FALCO O., THIANPRASIT M., KINT A. — Histologie et histochimie de lésions psoriasiques traitées par des pansements occlusifs à base de fluorandrenolone. *Bull. Soc. Franç. Derm. Syph.*, 1963, *70*, 235–238.

[17] BURROWS M. T. — The significance of the lunula of the nail. *Hohns Hopk. Hosp. Rep.*, 1919, *18*, 357–361.

[18] CONARD V., ACHTEN G. — Observation personnelle.

[19] DUPERRAT B., CINTRACT M. — Panaris mélanique. *Bull. Soc. franç. Derm. Syph.*, 1960, *7*, 235–242.

[20] DUPERRAT B., MASCARO J. M. — Melanomas subungéales. Estudio de ventium casos. *Actas Dermosif (Madrid)*, 1966, *57*, 5–18.

[21] ELLER J. J., ANDERSON N. P. — Leuconychia totalis clinical report with a review of the literature. *Med. J. Rec.*, 1928, *21*, 318–319.

[22] GADRAT J., BAZEX A., DUPRÉ A. — Réaction d'Hotchkiss MacManus comme méthode de détection en dermatologie. *Bull. Soc. franç. Derm. Syph.*, 1952, *59*, 375–377.

[23] HAM A. W., LEESON T. S. — *Histology*. 4th Edit., London, Pitman Medical, 1961.

[24] HASHIMOTO K., GROSS B. G., NELSON R., LEVER W. F. — The ultrastructure of the nail in 16–18 weeks old embryos. *J. invest. Derm.*, 1966, *47*, 205–217.

[25] HEIDINGSFELD M. L. — Leucopathia unguium. *J. cutan. Dis.*, 1900, *18*, 490.

[26] HIGASHI N. — Melanocytes of nail matrix and nail pigmentation. *Arch. Derm.*, 1968, *97*, 570–574.

[27] HIGASHI N., TADEO S. — Horizontal distribution of the DOPA + melanocytes, in the nail matrix. *J. invest. Derm.*, 1969, *53*, 163–165.

[28] JARRETT A., SPEARMAN R. I. C. — The histochemistry of the nail. *Arch. Derm.*, 1966, *94*, 652–657.

[29] JILLSON O. F., PIPER E. L. — The role of saprophytic fungi in the production of eczematous dermatitis. *J. invest. Derm.*, 1957, *28*, 137–146.

[30] KLIGMAN H., MESCON H., DELAMATER E. — The Hotchkiss-McManus stain for the histologic diagnosis of fungus disease. *Amer. J. clin. Path.*, 1951, *21*, 85.

[31] LEWIN K. — The normal finger nail. *Brit. J. Derm.*, 1965, *77*, 421–430.

[32] LEWIN K. — The finger nail in general disease. *Brit. J. Derm.*, 1965, *77*, 431–438.

[33] LEWIS B. L. — Microscopic studies of fetal and mature nail and surrounding tissues. *Arch. Derm. Syph. (Chic.)*, 1954, *70*, 732–747.

[34] MITCHELL J. C. — A clinical study of leuconychia. *Brit. J. Dermat.*, 1953, *65*, 221–230.

[35] PACK G. T., OROPEZA R. — Subungual melanoma. *Surg. Gynec. Obstet.*, 1967, *124*, 571–582.

[36] PARDO-CASTELLO V., PARDO O. A. — *Diseases of the nails*. Springfield, Charles C. Thomas, 1960.

[37] PILLSBURY D. M., SHELLEY W. B., KLIGMAN A. M. — *Dermatology*. Philadelphia, Saunders, 1956, 1021–1022.

[38] PRUNIERAS M. — PAB: A valuable stain for connective tissue, keratin and fungi. *J. invest. Derm.*, 1960, *35*, 309–314.

[39] PRUNIERAS M., DE BEER P. — Le PBA dans la coloration des champignons sur coupes et culture. *Bull. Soc. franç. Derm. Syph.*, 1960, *67*, 336–340.

[40] PRUNIERAS M., DE BEER P. — Le permanganate – bleu Alcian comme méthode de coloration de Candida albicans sur coupes et sur culture. *Bull. Soc. franç. Derm. Syph.*, 1960, *67*, 685–689.

[41] PRUNIERAS M., DE BEER P. — Relation entre l'épiderme et les champignons pathogènes. Application de la réaction au PBA. *Bull. Soc. fanç. Derm. Syph.*, 1961, *68*, 294–297.

[42] ROOBAERT N., ACHTEN G. — Etude «in vitro» de l'envahissement de la kératine unguéale par les dermatophytes. *Arch. belg. Derm. Syph.*, 1969, *25*, 121–123.

[43] ROTHBERG S., CROUNSE R. G., LEE J. L. — Glycine – C 14 incorporation into the proteins of normal stratum corneum and the abnormal stratum corneum of psoriasis. *J. invest. Derm. Syph.*, 1961, *37*, 497–504.

[44] SAMMAN P. D. — The ventral nail. *Arch. Derm.*, 1961, *84*, 192–195.

[45] SAMMAN P. D. — *The nail in disease*. London, William Heinemann, 1965.

[46] SIBLEY K. — Leuconychia striata. *Brit. J. Derm.*, 1922, *34*, 238.

[47] SINGER P. L. — Leukonychia: Its normal occurrences and causation. *Arch. Derm. Syph. (Chic.)*, 1931, *24*, 112–115.

[48] STONE O. J., MULLINS J. F. — Chronic paronychia. Microbiology and histopathology. *Arch. Derm. Syph.*, 1962, *86*, 324.

[49] TERRY R. — The onychodermal band in health and disease. *Lancet*, 1955, *1*, 179–181.

[50] ZAIAS N. — Embryology of the human nail. *Arch. Derm.*, 1963, *87*, 37–53.

[51] ZAIAS N. — The longitudinal nail biopsy. *J. invest. Derm. Syph.*, 1967, *49*, 406–408.

[52] ZAIAS N. — The movement of the nail bed. *J. invest. Derm. Syph.*, 1967, *48*, 402–403.

[53] ZAIAS N., ALVAREZ J. — The formation of the primate nail plate. An autoradiographic study in squirrel monkey. *J. invest. Derm.*, 1968, *51*, 120–136.

2. NAIL MALFORMATIONS

R. Malek

Nail malformations are not uncommon, yet often they are little understood by hand surgeons. This study may have little practical value, but it is intended to assist the interpretation of the pathological embryology of congenital malformations of the hand. There are four sections:

the embryology of the nail

the most common nail anomalies

a correlation of these anomalies of the nail with hand malformations

finally, the nail is frequently abnormal in the absence of any other hand malformation in a number of general malformation syndromes of which some examples will be given.

EMBRYOLOGY OF THE NAIL

The nail, like all exoskeletons, comes from the ectodermis. The formation of the nail is a continuous process and its several stages of development can only be separated artificially.

The nail starts to form in the first week of the third foetal month with the appearance of a transverse, dorsal ectodermal fold situated opposite the base of the terminal phalanx, which is already well formed (Fig. 2.1). The limits of the nail area are defined by the formation of a curved groove, concave proximally, and by a distal terminal groove which crosses the tip of the finger at an equal distance from the palmar and dorsal surfaces. The nail area then recedes to become completely dorsal during the fourth month. The ectodermis thickens and a first keratinised layer appears on the surface. This is called the

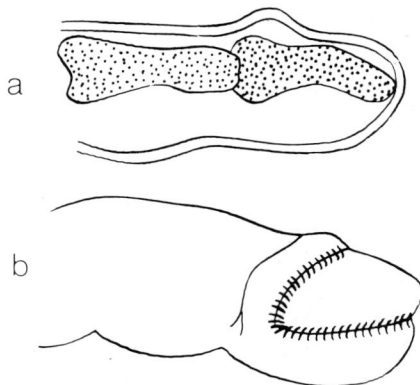

Figure 2.1
(a) The initial ectodermal fold (third month).
(b) Limits of the nail area and the terminal groove.

'eponychium' (Fig. 2.2). This eponychium then extends proximally while its middle part separates, leaving only the two ends. The proximal end, or 'epionyx', seems to penetrate the base of the dorsal fold and will help to form the nail root. A second keratinised layer then appears beneath the base of the eponychium and replaces it, becoming the primary nail. It looks irregular and friable and still a long way in appearance from the final nail. It starts to form in the second week of the fifth month and this time grows distally towards the fingertip.

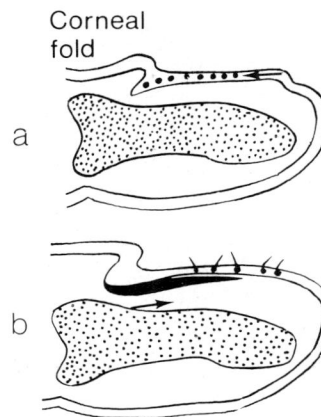

Figure 2.2
(a) Keratinised layer of eponychium and dorsal migration of the terminal groove.
(b) Appearance of the primary nail in the fourth month.

It is important to note that the slow, embryological development of the nail occurs at a stage when the differentiation of the mesenchymal tissues (bones, joints, muscles) is complete. Malformation of the nail can be related to arrested development, a primary lesion of the ectodermis, or it can be a consequence of an initial mesodermal anomaly affecting the development of the ectodermis.

NAIL ANOMALIES

Nail anomalies are numerous and have rather elaborate names to describe simple disorders:

anonychia, complete absence of the nail

onychoheterotopia, nail in an abnormal position (especially auricular)

onychatrophy, a small friable nail, often detached from its bed *pachyonychia*, large, hard, thick nails, often with abnormal striations.

Other anomalies which occur include pigmented striation, variation in shape, for example, nails that are wider than they are long, spherical nails, or cylindrical in appearance, etc.

It is important to note that a malformed nail, if removed, will regrow in identical form. Therefore, there is little point in removing such a nail for aesthetic reasons.

THE NAIL IN MALFORMATIONS OF THE HAND

Each of these anomalies can be observed in association with other malformations of the hand; sometimes several forms are found together in the same hand. In contrast, however, the nail may frequently be normal on a malformed finger (Fig. 2.3). This

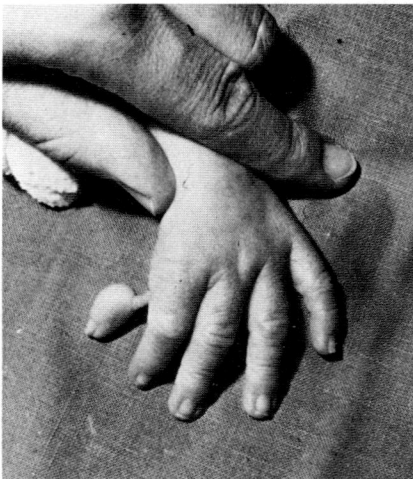

Figure 2.3
Healthy nail on a severed thumb.

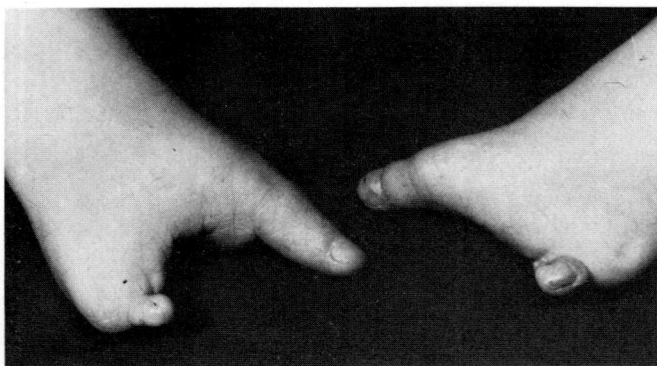

Figure 2.4
Nail anomalies in a lobster claw deformity.

is not surprising since, as has been described above, the nail does not develop at the same time as the other parts of the hand. If there is no anomaly of the nail, the teratogenic factor has ceased

to operate before the third month and, moreover, it has not involved the ectodermis.

However, if the development of the ectodermis is faulty as a result of a mesodermal anomaly which occurred earlier, a nail deformity may develop, as in the following circumstances:

(a) in the gross abnormalities, such as lobster-claw hand or microdactyly, the nail is short, convex and cylindrical (Fig. 2.4)
(b) in amniotic disease, there may be persistence of actual embryonic nail areas with an eponychium (Fig. 2.5).

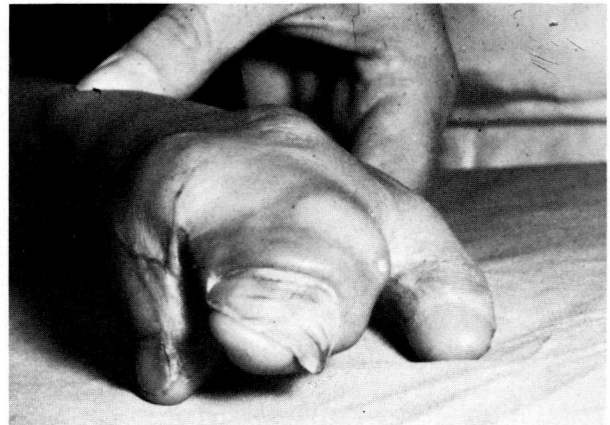

Figure 2.5
Case of amniotic disease with embryonic nail areas.

This would suggest that amniotic disease appears late, yet it must originate earlier to explain the bone anomalies, which occur so frequently. The nail is often completely absent in amniotic disease. Congenital amputation has been suggested as the explanation (Fig. 2.7). However, this may not be so for we have seen in such finger stumps the frequent occurrence of a terminal furrow which is similar to the terminal groove (Fig. 2.8).

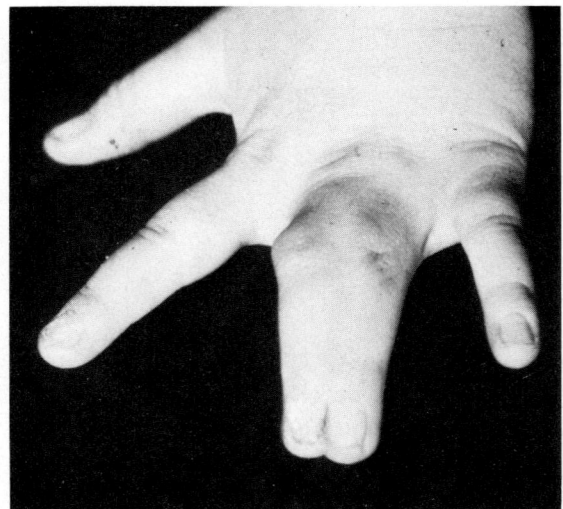

Figure 2.6
Single nail separated by a furrow in syndactyly with terminal synostosis.

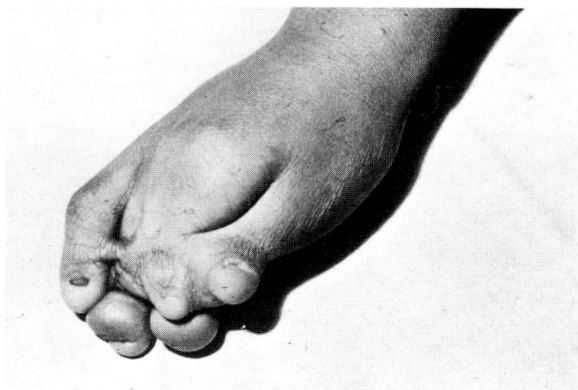

Figure 2.7
Amniotic disease with a variety of nail anomalies.

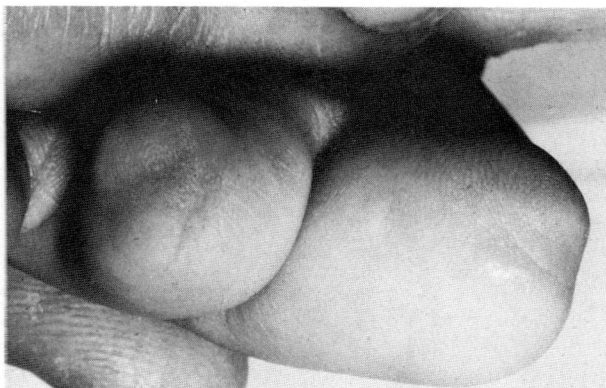

Figure 2.8
Beginnings of a nail on finger stumps in amniotic disease.

1. In congenital ectodermal dysplasias the nail is often missing:
    (a) congenital ectodermal defect syndrome in which striated and friable nails are associated with alopecia and the absence of sweat and sebaceous glands
    (b) the nail is sometimes absent in ichthyosis
    (c) congenital dyskeratoses are accompanied by pachyonychia

2. The nail is often malformed in craniocleidodysostosis and the trisomies.

3. In Crouzon's disease and Apert's syndrome the nail is frequently thick and wide with a single fused nail present on the multiple syndactylies, which are a feature of these conditions (Fig. 2.9).

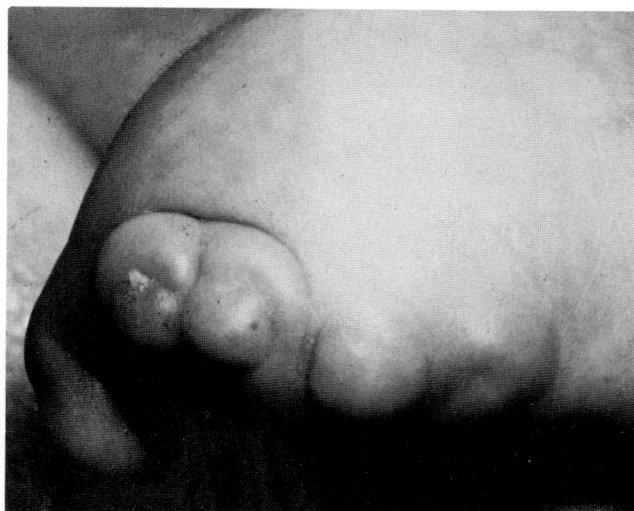

Figure 2.9
Nail in Apert's syndrome.

In type A syndactyly, where the fusion involves the terminal phalanges, there is often just one nail, sometimes divided longitudinally by a furrow (Fig. 2.6). After the nail has been cut, a full-thickness skin graft applied to the lateral surfaces of the finger must be carefully sutured to the epidermis of the nail bed. In the months following this operation, a healthy nail fold develops by thickening and migration of the soft part; this occurs even when the graft has been applied to a bone surface in the case of syndactyly with synostosis of the terminal phalanges.

In other types of malformation of the hand, which would take too long to describe, the nail may be malformed, but more frequently it is of normal appearance.

SYNDROMES WITH WIDESPREAD MALFORMATION

The nail is frequently malformed in syndromes involving congenital abnormalities in tissues apart from the hand. There are many such cases in which the nail anomaly is really part of a collection of characteristic features. Sometimes, however, it is the anomaly that indicates the diagnosis.

4. In mongolism the nails are rectangular.

Finally, a syndrome familiar to orthopaedic surgeons as the nail patella syndrome, or osteo-onychodysplasia (Figs. 2.10 and

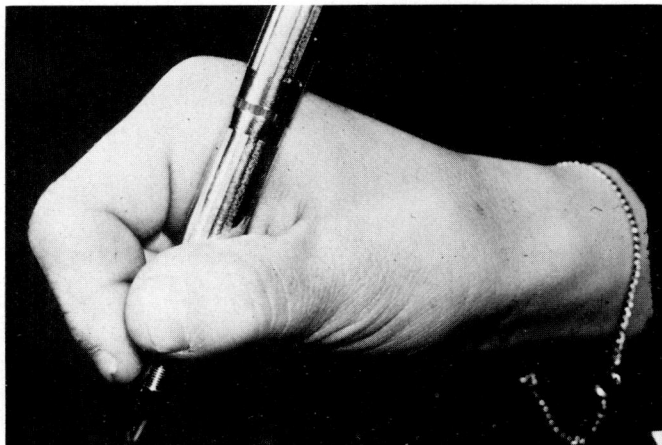

Figure 2.10
Nail patella syndrome.

2.11). In this syndrome the nail anomaly is often confined to the thumb and index finger and is associated with dislocation of the patellae and the presence of curious iliac horns, which can be felt and are visible on X-rays of the pelvis.

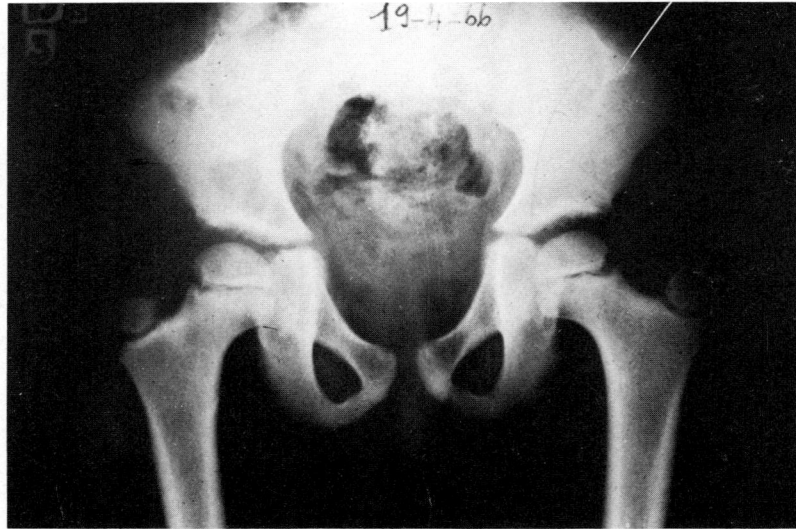

Figure 2.11
Nail patella syndrome.

3. PRINCIPAL MODIFICATIONS OF THE NORMAL FORM OF THE NAIL

R. Baran

Understanding of nail pathology is hampered by its limited symptoms and the subtlety required in decoding. However, the use of modern research techniques (such as electron microscopy), the introduction of powerful fungicides, and the use of surgery for biopsy and treatment are advancing diagnosis and treatment and justify special consideration of nail disorders within the field of medicine as a whole.

CLUBBING OR HIPPOCRATIC NAIL

The increased nail curvature which characterises this disorder affects all the nails, but predominantly the radial three digits (Table 3.1), and it can be of three types: longitudinal, the hooked nail, like a bird's beak (Fig. 3.1); transverse, shaped like the balance wheel of a watch or a snake's head; a mixed type, like a watch glass. There are four main categories of finger clubbing.

TABLE 3.1 Aetiological classification of clubbing
_____

Idiopathic forms

Hereditary and congenital forms sometimes associated with other anomalies:

      Familial and genotypic pachydermo-periostosis
      Racial forms (Negroes, North Africans)

Acquired forms:
1. The thoracic organs are involved in about 80 per cent of cases of clubbing, often with the common denominator of hypoxia:
   - (a) Broncho-pulmonary diseases, especially chronic and infective
     Bronchiectasis, abscess and cyst of the lung, pulmonary tuberculosis
     Sarcoidosis, pulmonary fibrosis, emphysema, Ayerza's syndrome, chronic passive congestion
     Blastomycosis, pneumonia
   - (b) Thoracic tumours
     Primary or metastatic broncho-pulmonary cancers, pleural tumours
     Mediastinal tumours (infrequently a cause), Hodgkin's disease, lymphoma, pseudo-tumour due to oesophageal dilatation
   - (c) Cardio-vascular diseases
     Congenital heart disease associated with cyanosis
     Thoracic vascular malformations: stenoses and arterio-venous aneurysms
     Subacute bacterial endocarditis
     Congestive cardiac failure
     Myxoma
     Raynaud's syndrome, erythromelalgia, Maffucci's syndrome

2. The gastro-intestinal forms represent less than 5 per cent of cases:
   - (a) Oesophageal, gastric and colonic cancers
   - (b) Diseases of the small intestine
   - (c) Colonic diseases with:
     amoebiasis and inflammatory states of the colon
     ulcerative colitis
     familial polyposis, Gardner's syndrome
     ascariasis, whipworm infestation.
   - (d) Active chronic hepatitis
     Primary or secondary cirrhoses

3. Endocrine origin:
   Diamond's syndrome (pretibial myxoedema, exophthalmos and finger clubbing)

(continued on next page)

TABLE 3.1 (*continued*)

4. Haematological:
    Primary polycythaemia or secondary polycythaemia associated with hypoxia
    Poisoning by phosphorus, arsenic, alcohol, mercury or beryllium
5. Hypervitaminosis A
6. Neural: Syringomyelia
7. Unilateral or limited to a few digits:
    Subluxation of the shoulder (paralysis of the brachial plexus)
    Median neuritis
    Pancoast-Tobias syndrome
    Aneurysm of the aorta or the subclavian artery
    Sarcoidosis
    Tophaceous gout
8. Isolated forms:
    Local injury, whitlow, lymphangitis
    Subungual epidermoid inclusions
9. Transitory form: physiological in the newborn child (due to reversal of the circulation at birth)
10. Occupational acro-osteolysis (exposure to vinyl chloride)

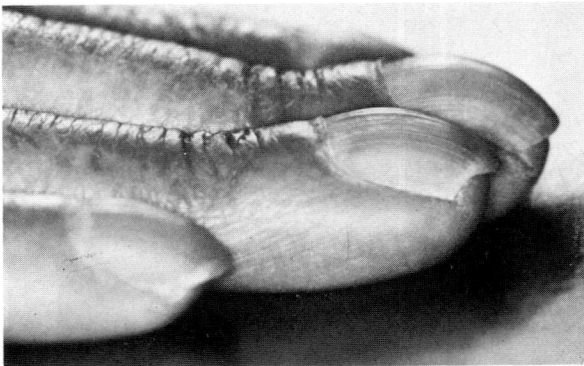

Figure 3.1
Clubbing or hippocratic nail.

1. *The simple type* which is the most common. It has several elements:
  (a) increased nail curvature. When this is beginning or receding, a furrow separates it from the rest of the nail
  (b) hypertrophy of the soft parts of the terminal segment due to a firm, elastic, oedematous infiltration of the pulp, which can spread on to the dorsal surface forming a periungual swelling: hyperplasia of the dermal fibro-vascular tissue explains why the matrix area is so readily involved
  (c) inconstant local cyanosis.

Coury and Saxe (1969) distinguish four stages of clubbing: suspected, slight, average and severe. In practice, clubbing is defined as occurring as soon as the proximal nail fold makes an angle of more than 180° with the nail plate (Lovibond's angle).

Radiological changes occur in less than a fifth of cases and include phalangeal demineralisation, an irregular thickening of the cortical diaphysis and an incipient periosteal cuff around the distal third of the proximal phalanx.

The pathological process, which appears to be responsible for clubbing and its associated changes, is hypervascularity resulting from the opening of many anastomotic shunts (Ponchon *et al.*, 1969).

2. *Hypertrophic osteoarthropathy* is a rare disorder associated with proliferative periostosis and often rarifying osteitis of the phalanges and bones of the forearm. It is accompanied by pseudo-inflammatory, symmetrical arthropathies of the major joints of the limbs, in particular the lower limbs. This syndrome is practically pathognomonic of malignant broncho-pulmonary neoplasm.

3. *Pachydermoperiostitis* is very rare. The lesions of the fingertips are clinically identical to those of hypertrophic pulmonary osteoarthropathy. However, in the latter condition, X-rays may show a translucent line between the periosteal new bone formation and the original bone, whereas in pachydermoperiostitis the periosteal apposition blends into the underlying bone without a line of demarcation, and decalcified areas do not occur. The pachydermal change of the extremities and face is the most characteristic feature of this disorder, which is referred to in French literature as the Touraine, Solente, and Golé syndrome.

4. *The shell nail syndrome*, only recently described (Cornelius and Shelley, 1967), occurs in some cases of bronchectasis and is similar to clubbing but with atrophy of the nail bed and the underlying bone.

KOILONYCHIA

Koilonychia is the opposite of clubbing as the nail is concave, the so-called 'spoon nail' (Baran and Achten, 1969; Bergeron and Stone, 1967). The underlying tissues may be healthy or affected by subungual hyperkeratoses, which are clearly visible at the margins indicating an external origin of the deformity. The nail, which may be normal, thinned or thickened, has a smooth surface. The nails may be soft. This dystrophy normally affects several fingers, sometimes all, but it affects the toes less often.

The petaloid nail is a variant of koilonychia at an early stage with flattening of the nail as its characteristic.

TABLE 3.2 Aetiological classification of koilonychia (modified from Bergeron and Stone, 1967)

Idiopathic forms

Hereditary and congenital forms, sometimes occurring with other anomalies:
    Fissured nails, sebocystomatosis
    Monilethrex, keratoderma of the palm (Meleda type), leukonychia
    Nail-patella syndrome
    Nezelof's syndrome (immunological defect)

Acquired forms:
1. Cardio-vascular and haematological:
    Iron deficiency anaemia (following gastrectomy, Plummer-Vinson syndrome)
    Iron malabsorption by the intestinal mucosa
    Haemoglobinopathy SG
    Polycythaemia
    Haemochromatoses
    Banti's syndrome (the nails heal after splenectomy)
    Coronary disease
2. Infections: syphilis; fungal diseases
3. Endocrine forms:
    Acromegaly
    Hypothyroidism
    Thyrotoxicosis
4. Traumatic and occupational forms:
    Petrol, various solvents, engine oils
    Acids and alkalis, thioglycolate (hairdressers)
    Housewives, chimney sweeps, rickshaw men (attacks the feet)
    Nail biting
5. Avitaminoses (P, B2 and especially C)
    Cystine deficiency
6. Dermatoses: Raynaud's disease, scleroderma, lichen planus, acanthosis nigricans, porphyria cutanea tarda, Darier's disease, incontinentia pigmenti, alopecia areata
7. Toe-nails: found in certain young children who are otherwise normal
8. Kidney transplantation

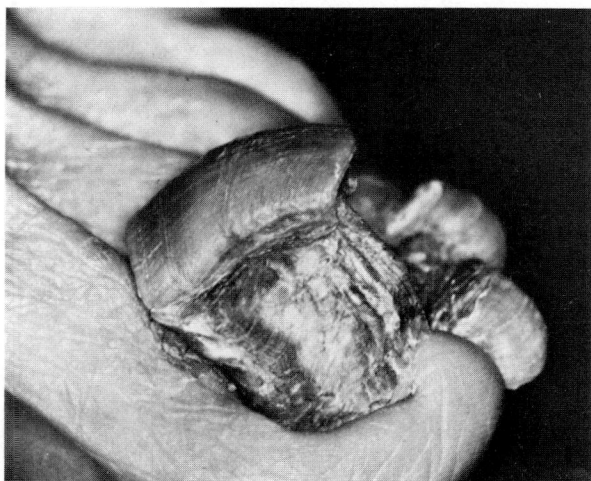

Figure 3.2
Onychauxis.

THICKENED NAILS (HYPERONYCHIA)

Onychauxis is a regular thickening of the nail plate, which may become stratified (Fig. 3.2). The term pachyonychia covers both this variety as well as thickening by hyperkeratosis of the subungual tissues, principally the hyponychium, causing the distal part of the nail to be elevated by a friable accumulation, which causes the nail to become loosened from its bed.

The term pachyonychia congenita (Fig. 3.3) describes a simple or complex dystrophy, which is the major feature of Jadassohn-Lewandowsky's syndrome. The nails are yellow–brown in colour and extremely hard. They are barrel-shaped, the nail plate having an exaggerated transverse curvature with its free edge shaped like a horseshoe. All the nails are affected but the toe-nails less severely. They remain firmly adherent to their bed. Recurrent paronychial inflammation is possibly responsible in certain cases for repeated shedding of the nails.

Onychogryphosis (Fig. 3.4) is a ram's horn nail. It occurs principally in the toes, particularly in the hallux, and is due both

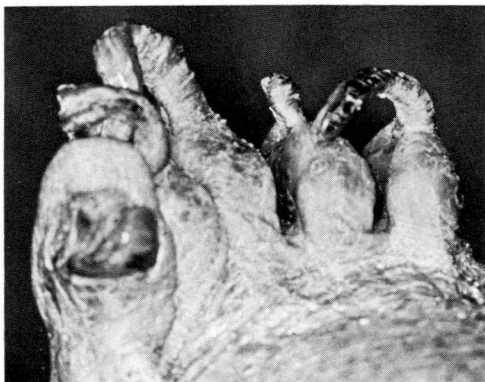

Figure 3.3
Pachyonychia congenita

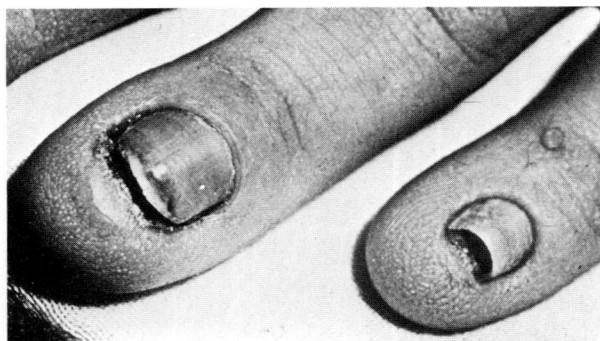

Figure 3.4
Onychogryphosis.

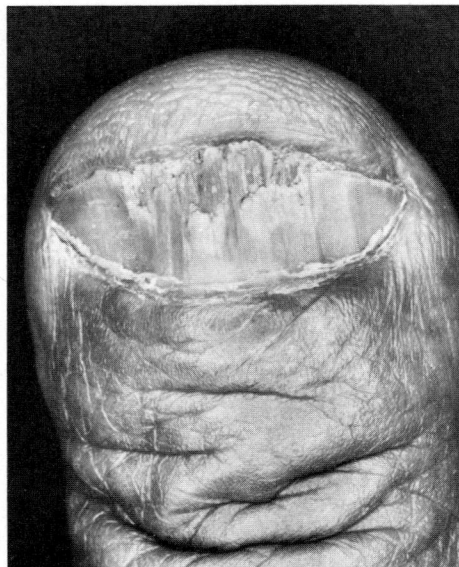

Figure 3.5
Racquet thumb.

recently described the syndrome of 'broad thumbs and toes and facial anomalies'.

INCREASED TRANSVERSE CURVATURE OF THE NAIL

There are two main types of this condition: the arched, pincer or trumpet nail; and the tile-shaped nail with lateral folds.
1. The arched nail is a dystrophy characterised by increased transverse curvature, which becomes more marked towards the tip. The long axis of the nail is increased.

The pincer nail is a special kind of ingrowing nail – the lateral edges squeeze the soft tissues but without necessarily breaking the epidermis. Eventually these tissues atrophy and this may be accompanied by resorption of the underlying bone (Cornelius and Shelley, 1968). Other cases may be due to a subungual exostosis which requires to be removed. Like a clamp, the lateral grooves maintain a permanent deformation of the nail plate. At its most marked the edges come together forming a tunnel, or coil up like a trumpet (Baran, 1974, 1975) (Fig. 3.6).

In certain varieties the nails are claw-shaped or resemble pachyonychia congenita (Samman, 1978). These anomalies would be a minor disfigurement but in some cases they become painful from pressure, even by the weight of a sheet in bed, and relieved only by avulsion of the nail. Recurrence is not unusual and, to prevent this, surgical destruction of the matrix may be required.
2. The tile-shaped nail is characterised by increased transverse curvature but in this condition the lateral borders of nail remain parallel.

In the instance of plicated nail the central part of the nail is more or less flat, while one or both margins are sharply angled forming

to a hyperplasia of the nail bed and thickening of the nail plate; the latter has transverse striations and is brownish, opaque, and spiral-shaped with an irregular surface.

Two varieties can be distinguished. The first is common among young people but can occur at all ages and results from injury involving the matrix, which produces thickening of the germinal layer. In the second, which occurs in old age, subungual keratin pushes up the nail plate. A combination of the two types occurs in fungal onychogryphosis. It occurs usually in people who do not take good care of themselves. Although this dystrophy may be a source of discomfort, or even pain, when shoes are worn, it is usually accepted without complaint.

The knowledge of the 'congenital malalignment of the big toe nail' and its treatment help to prevent later problems such as hemi-onychogryphosis.

SHORT NAILS (BRACHYONYCHIA)

This term is applied to nails that are shortened and wider than they are long. It may occur on its own or together with a shortening of the terminal phalanx, creating, for example, the 'racquet thumb' (Fig. 3.5). The condition is usually inherited as an autosomal dominant. Rubinstein and Taybi (1963) have

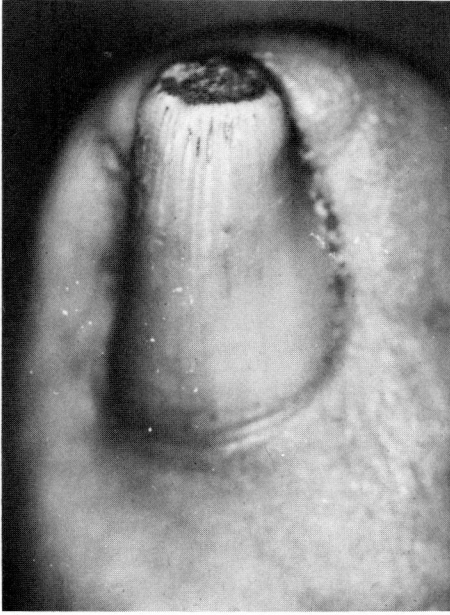

Figure 3.6
Trumpet nail.

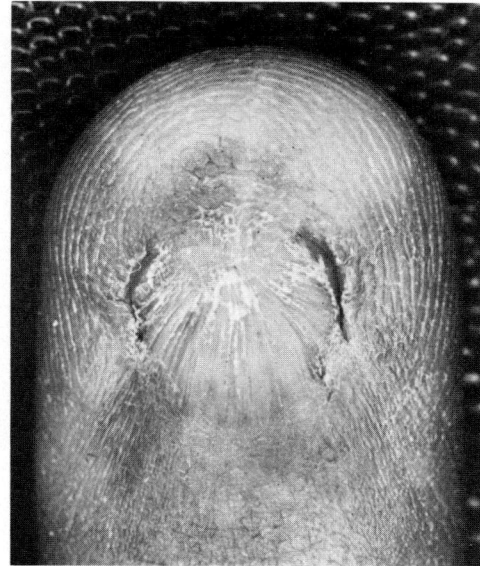

Figure 3.7
Lichen planus onychatrophy.

vertical sides which are parallel. These deformities may be associated with ingrowing nails in which the compression of soft parts produces inflammatory oedema and a vicious circle is established in which the deformed nail and the soft parts are in painful conflict.

WORN NAILS

Wear of the free edge of the nails can be due to excessive scratching in cases of chronic pruritis (this also polishes the surface). It is seen also in many manual workers. Together with Schubert *et al.*, we have recently described this condition as an occupational hazard of mushroom-pickers caused by handling plastic bags.

NAIL ATROPHY

Acquired or congenital atrophy of the nails is a reduction in size and thickness of the nail plate, often associated with fragmentation and fissuring. In its most complete form, a scar replaces the absent nail (Fig. 3.7).

ANONYCHIA

This implies absence of all or part of the nails. There are three types:
(a) aplastic (the nail has never formed)
(b) atrophic (the rudimentary nail is reduced to a corneal layer)

(c) hyperkeratotic (the nail is replaced by a thicker layer of keratin).

Anonychia may be found on its own, usually transmitted as a dominant, or may be part of a more widespread syndrome.

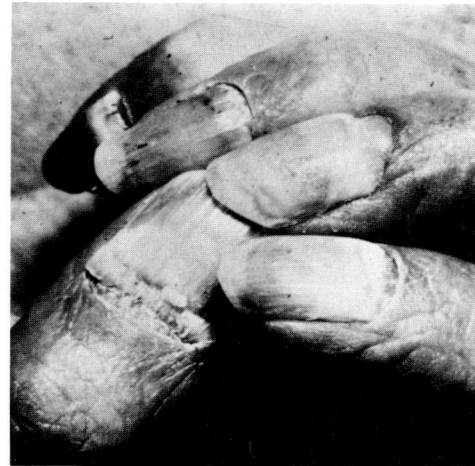

Figure 3.8
Onychomadesis in Lyell's syndrome (Puissant's collection).

NAIL SHEDDING

Spontaneous separation of the nail from the matrix is called onychomadesis (Fig. 3.8). It is due to a limited lesion of the proximal part of the matrix. Total loss of the nail may also result from onycholysis (Fig. 3.9).

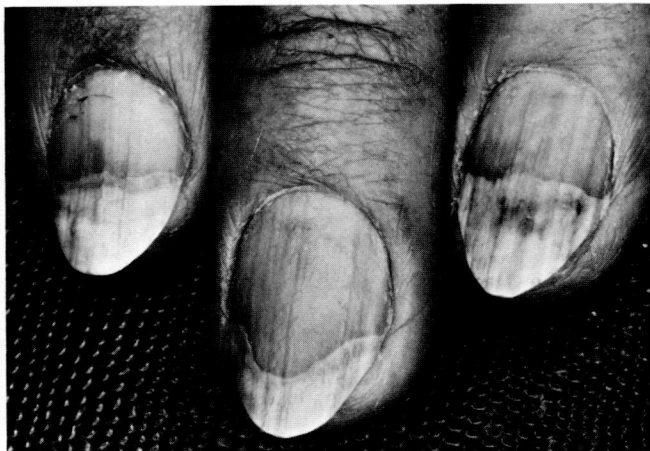

Figure 3.9
Thyrotoxic onycholysis.

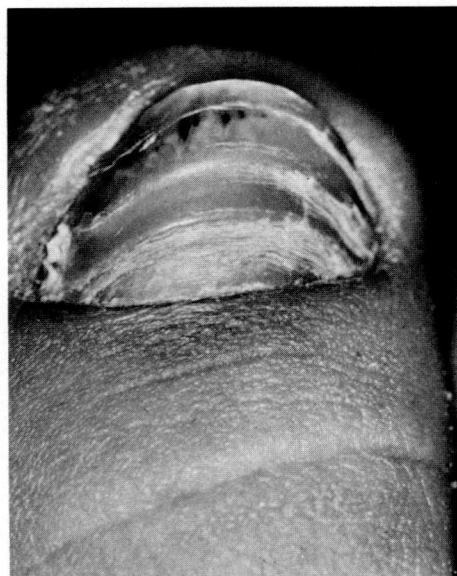

Figure 3.10
Tennis toe with transverse grooves and distal splinter haemorrhages.

The terms onychoptosis, defluvium, or alopecia ungium are sometimes used to describe such nail loss. When the disease is inherited (as a dominant), the shedding may be periodic. Separation of the nail can also be due to injury which may be quite minor or repetitive as, for example, in sportsman's toe.

SPORTSMAN'S TOE (TENNIS TOE)

This is a problem of sports enthusiasts (skiers, climbers, tennis and football players). The toe suffers haemorrhages, first in the matrix, then under the nail, and these cause nail separation, due to a combination of onychomadesis and onycholysis. Normally, it is the longest toe that is affected, but both the first and the second toes may be involved simultaneously (Fig. 3.10) and bilaterally.

Encysted subungual haematomas are liable to become infected, unlike the splinter haemorrhages which occur with equal frequency. During regrowth of the nail, certain complications may occur, such as subungual exostosis, ingrowing of the nail, sometimes surrounded by an anterior wall of soft tissue, and various irregularities of the nail itself.

ONYCHOLYSIS

Onycholysis refers to the detachment of the nail from its bed at its distal end and/or its lateral attachments (Baran and Temine, 1973). The detachment extends proximally along a convex line,

giving the appearance of a half-moon. When the process reaches the matrix, onycholysis becomes complete. Involvement of the lateral edge of the nail plate alone is unusual (Fig. 3.9). In certain cases, the free edge rises up like a hood or coils up on itself like a roll of paper.

Onycholysis creates a space under the nail which is usually filled with keratin debris; the greyish-white colouring is due to the presence of air under the nail.

In onycholysis due to pachyonychia there is a marked thickening of the subungual keratin which elevates the nail plate. Histology reveals PAS positive, homogeneous, rounded or oval-shaped, amorphous masses surrounded by healthy squamous cells usually separated from each other by empty spaces. The colour varies from yellow to brown, depending on aetiology; the area is rarely malodorous.

Onycholysis is usually symptomless and it is mainly the appearance of the nail that brings the patient to the doctor. Sometimes, however, there may be slight pain associated with inflammation at an early stage of its development.

The extent of onycholysis may be estimated by measuring the distance separating the distal edge of the lunula from the limit of proximal detachment (Taft, 1968). Transillumination of the terminal phalanx gives a good picture of the area.

The various causes which can produce onycholysis are listed in Table 3.3.

TABLE 3.3 Classification of Onycholyses (modified from Ray, 1963)

Idiopathic

Systematic

1. Circulatory
2. Endocrine (hypothyroidism, thyrotoxicosis, etc.)
3. Pregnancy
4. Syphilis
5. Iron deficiency anaemia

Cutaneous diseases
1. Psoriasis
2. Contact dermatitis – atopic dermatitis, escavenitis (Nesreis diversicolour)
3. Hyperhidrosis
4. Congenital onycholysis, partial hereditary onycholysis, onycholysis due to pachyonychia
5. Various drug-induced eruptions, bleomycin, doxorubicin, aromatic retinoids
6. Treatments inducing photosensitivity: cyclines (especially demethyl-chlortetracycline and doxycycline), trypoflavine, chlorpromazine, psoralenes with U-V A or sunlight

Local causes
1. Traumatic causes (accidental, occupation, or mixed)
2. Infectious causes:
fungal
bacterial
viral
3. Chemical causes:
hot water with alkalis and/or detergents, sodium hypochlorite
petrol and various solvents, anti-rust agents
sugar solutions
cosmetics (base coats, hardeners such as formaldehyde, false nails, hair setters and depilatory products)
4. Physical causes:
burns

BIBLIOGRAPHY

[1] BARAN R. — Hypercourbure unguéale transverse. Ongle en pince. Ongle en cornet. *Bull. Soc. fr. Derm. syph.*, 1975, *82*, 55–56.
[1c] BARAN R. — Pincer and trumpet nails. *Arch. Derm.*, 1974, *110*, 649.
[2] BARAN R., ACHTEN G. — Les associations congénitales de koïlonychie et de leuconychie totale. *Arch. belge Derm. Syph.*, 1969, *25*, 13–29.
[3] BARAN R., TEMIME P. — Les onychodystrophies toxicomédicamenteuses et les onycholyses. *Concours méd.*, 1973, *95*, 1007–1023.
[4] BERGERON J.-R., STONE O.-J. — Koilonychia. *Arch. Derm. (Chic.)*, 1967, *95*, 351–353.
[5] CORNELIUS C.-E., SHELLEY W.-B. — Pincer nail syndrome. *Arch. Surg.*, 1968, *96*, 321–322.

[6] CORNELIUS C.-E., SHELLEY W.-B. — Shell nail syndrome associated with bronchiectasis. *Arch. Derm. (Chic.)*, 1967, *96*, 694–695.
[7] COURY C. E., SAXE H. de. — L'hippocratisme didigital. *Gaz. méd. Fr.*, 1969, *76*, 5495–5508.
[8] PONCHON Y., CHELLOUL N., ROUJEAU J. — Contribution à l'étude anatomopathologique de l'hippocratisme digital. *Sem. Hôp. Paris*, 1969, *42*, 2604–2611.
[9] RAY L.-F. — Onycholysis. A classification and study. *Arch. Derm.*, *(Chic.)*, 1963, *88*, 181–185.
[10] RUBINSTEIN J.-H., TAYBI H. — Broad thumbs and toes and facial abnormalities. *Amer. J. Dis. Child*, 1963, *105*, 588.
[11] SAMMAN P. — *The nails in diseases*. London, William Heinemann, Medical Books, 1978, 3e edit.
[12] TAFT E.-H. — Onycholysis. *Aust. J. Derm.*, 1968, *9*, 345–351.

4. MODIFICATIONS OF THE NAIL SURFACE

R. Baran

LONGITUDINAL LINES

Longitudinal lines or striations can be indented (grooves) or projecting (ridges) (Fig. 4.1).

Figure 4.1
Koilonychia with strongly marked longitudinal lines and fissuring of the free edge.

GROOVES

These are longitudinal furrows, usually parallel, shallow and delicate, and separated by low projecting ridges. They become more prominent with age and in certain pathological states, such as lichen planus, rheumatoid arthritis, peripheral circulatory disorders, genetic anomalies, etc.

Onychorrhexis is a series of small longitudinal, parallel furrows on the nail surface, which have the appearance of being scratched by an awl or by sandpaper. Fissuring of the free edge is common (Fig. 4.1).

Local lesions, such as a mucous cyst or wart, may produce pressure on the nail matrix resulting in a wide, deep, longitudinal groove or canal which will disappear if the cause is removed.

Dystrophia mediana canaliform, described by Heller, is a transient, but recurrent form (Fig. 4.2); it is usually symmetrical and most commonly involves the thumbs. It may develop into a deep fissure which splits the free edge.

A similar condition is Leclercq's chevron-shaped, median, ungual dystrophy (naevus striatus symetricus unguis) (Leclercq, 1964), in which oblique grooves are formed about the midline of the nail. The condition may be familial, traumatic or idiopathic.

Pterygium. The proximal nail fold appears thinned and the cuticle extends over the nail, to which it becomes adherent. The nail plate is fissured and is progressively destroyed. This fusion of the cuticle, matrix and bed may be congenital, but is particularly characteristic of vaso-motor ischaemia and above all

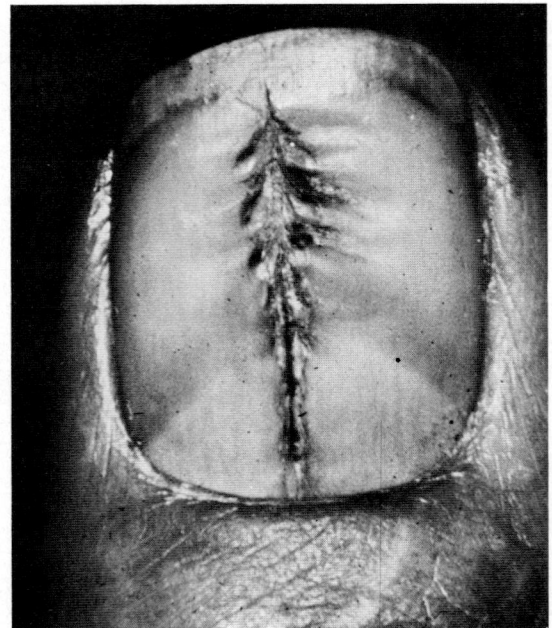

Figure 4.2
Heller's dystrophy (Leclercq's variety) (Achten's collection).

lichen planus (Fig. 4.3). It also occurs after radiotherapy and it may follow injury.

Ventral pterygium or ptergyium inversum unguis (Caputo & Prandi, 1973) is a distal extension of the hyponychial tissue that adheres to the under surface of the nail. It obliterates the distal furrow. It may be idiopathic but a similar appearance occurs in acrosclerosis, disseminated lupus erythematosus, and causalgia of the median nerve. Odom *et al.* (1974) described a 'congenital, aberrant, painful hyponychium', and there is a familial form of the disease (Dugois *et al.*, 1975).

LONGITUDINAL RIDGES

These small rectilinear projections extend as far as the free edge of the nail, or sometimes stop short. Sometimes a wide longitudinal median ridge has the appearance of a cross-section

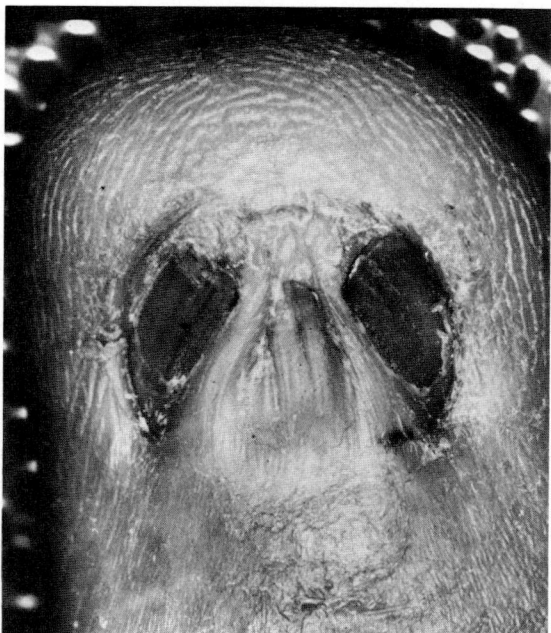

Figure 4.3
Pterygium due to lichen planus.

of a circumflex accent. The condition is inherited and affects mainly the thumb and index fingers of both hands. The longitudinal ridges may be interrupted at regular intervals (Fig. 4.4) giving a beaded appearance.

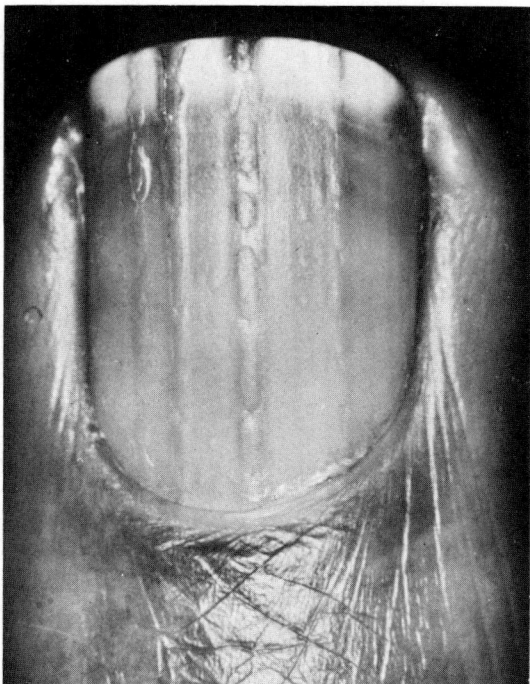

Figure 4.4
Accentuated longitudinal ridges resembling a string of sausages.

1. BEAU'S LINES — TRANSVERSE GROOVES

These are transverse linear grooves formed under the cuticle as a result of damage to the matrix. When the cause is a systemic disease (usually a severe febrile illness) all the nails of the fingers and toes are marked in the same place; these are called Beau's lines (Figs. 4.5 and 4.6). The condition is sometimes restricted to

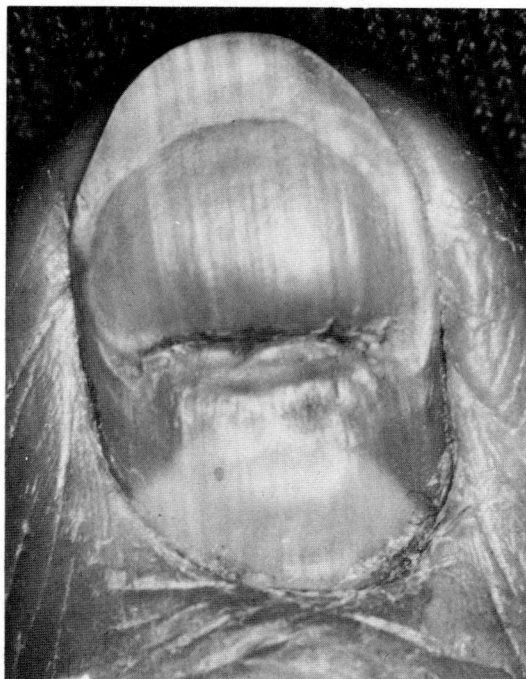

Figure 4.5
Beau's line.

the thumbs and big toes. The grooves are superficial, but more marked on the medial part of the nail and may extend almost through the full thickness. In severe cases the nail may become detached. The width of the groove shows how long the matrix has been affected since the nail grows 1 mm in every eight to ten days. The lines indicate that the germinal function of the matrix has been diminished.

Beau's lines enable a precise dating of the disease which caused them. A series of lines may be produced in cases of recurrent illness. Physiological Beau's lines may occur in five-week-old babies.

When the transverse furrows are the consequence of a local chronic condition, such as paronychia, eczema, or repeated injury due to over-zealous manicure, they are often numerous and curvilinear, like rippling. The grooves are separated by ridges of healthy nail.

A nervous habit of pushing back the cuticle repeatedly on one or several fingers, usually the index finger pushing back the cuticle of the thumb on the same hand, can result in a series of transverse grooves, which form a central longitudinal depression

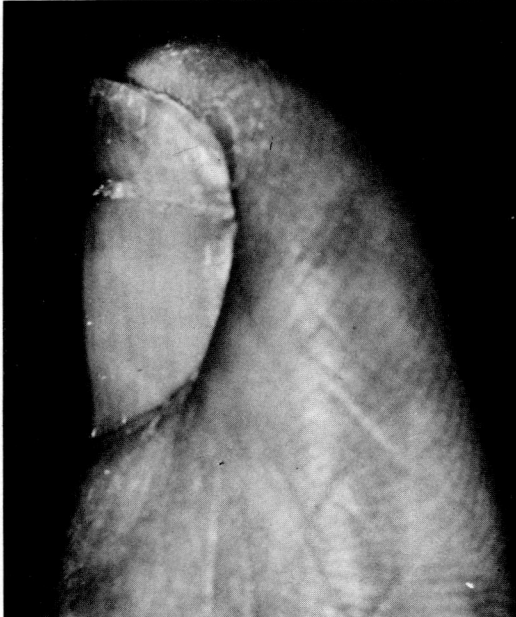

Figure 4.6
Beau's line.

on the nail, about 2 to 3 mm in width (Fig. 4.7) (washboard nails).

NEUROTIC NAIL DAMAGE

This disease takes three forms.

(a) The first results from picking at the nails or the surrounding area, producing loss of nail or hangnail. Similar appearances may result from nail-biting. Traumatic erosion of the nail folds may produce tags of epidermis, which are painful and susceptible to secondary infection.

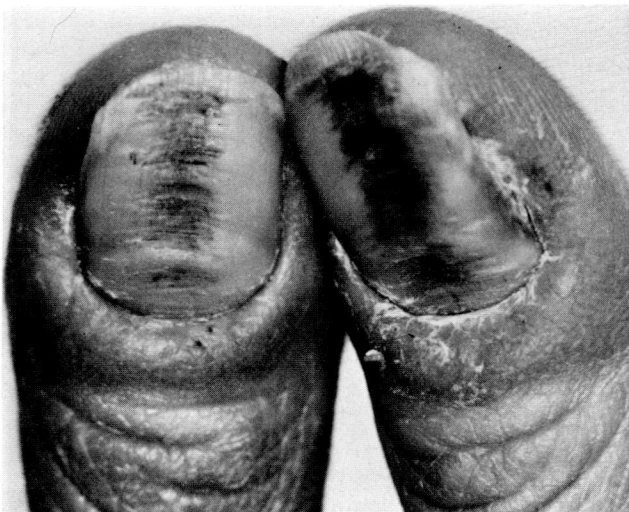

Figure 4.7
Neurotic pushing back of the cuticle.

(b) The second form is excessive pushing back of the cuticle, described above.
(c) The third is nail-biting, which can reduce the nail to a remnant that may become almost completely covered by a distal pulpy fold of soft tissue.

PITTING—ROSENAU'S DEPRESSIONS—PUNCTATE EROSIONS

The surface of the nail plate is covered by a varying number of depressions, the size of a pinhead, which may cover the entire nail (the thimble nail) (Fig. 4.8); or they can be arranged in series along one of several longitudinal lines; or they may be transversely-orientated and arranged in series like tide-marks.

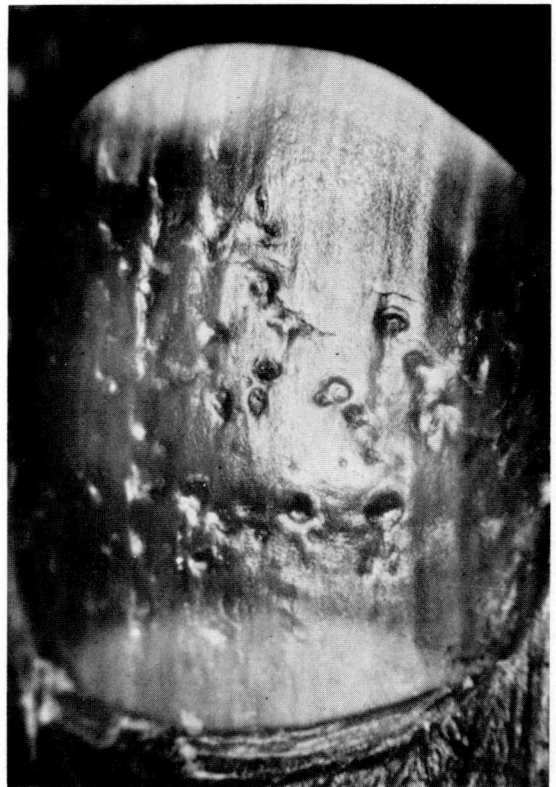

Figure 4.8
Pitting in psoriasis.

TRACHYONYCHIA

This dystrophy has been described by Alkiewicz (1950), and studied by Achten and Wanet-Rouard (1974), especially in relation to congenital nail atrophies.

It is characterised by roughness of the nail surface and a grey opacity of the nail. One form may result from external chemical action and the other is idiopathic, which may be congenital or acquired. The acquired form may be associated with a known dermatological disorder, such as lichen planus, psoriasis, or alopecia areata.

In its complete form trachyonychia can affect all the "sand papered nails" and it may even be a precursor of alopecia areata without, however, affecting the course of this disorder (Baran and Dupre, 1977).

NAIL FRAGILITY

This is a frequent complaint of our patients. There are three types which may occur singly or together.

SOFT NAILS

These may be thinner than usual. In hapalonychia the nail plate is particularly flexible and bluish in colour, semi-transparent and egg-shell; the distal edge is frequently cracked.

BRITTLE NAILS

These may be subdivided into four types.
(a) An isolated splitting of the free edge, sometimes extending proximally, as a longitudinal striation on the superficial layer of the nail with sometimes a dusty blackish discolouration (onychorrhexis).
(b) More commonly the splitting is multiple and may resemble the teeth of a comb or the battlements of a castle. Triangular fragments may be torn off the free edge if the nail catches on anything.
(c) Lamination of the free edge to a series of fine layers produces a scale-like lesion referred to as lamellar onychoschizia, which can be explained by the arrangement of the dorsal and the ventral nail plates (Fig. 4.9).
(d) Transverse cracks may develop which can become complete at the distal margin.

FRIABLE NAILS

Friability is sometimes confined to the surface of the plate, as in superficial white onychomycosis, or after the application of nail varnish (granulation of nail keratin). Friability may extend throughout the nail in some cases of psoriasis and in advanced fungal infections, particularly those due to dermatophytes. Nail friability is due to failure of one or more of the six factors on which the health of the nail depends: water content; chemical and physical injuries to the keratin constituent; composition of the nail plate; normal anatomy of the nail apparatus; genetic disorders; central and peripheral nervous system.

Local and systemic causes may be distinguished.

Local causes. Occupational and traumatic damage due to chemical agents, detergents, alkalis, solvents, sugar solutions, especially in hot water, may cause nail fragility. When one considers that the superficial layers of the skin are replaced within a month, while the nail is replaced in five or six months, it is easy to understand why the housewife, who exposes her hands to so many irritants during the course of everyday activities, is so vulnerable to nail complaints.

Fragility, due to nail cosmetics, is rare. Nails may be affected

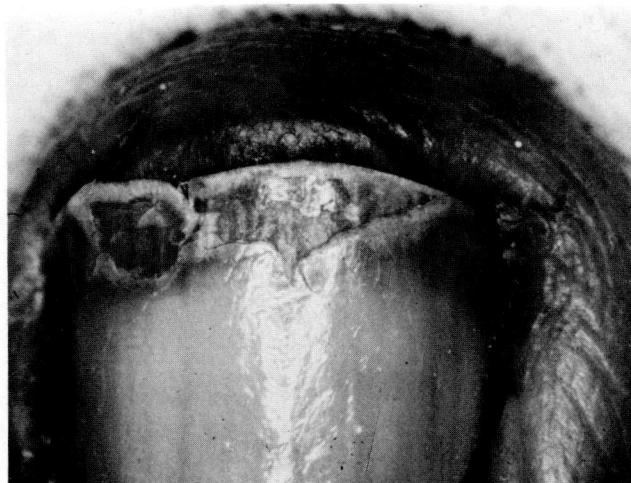

Figure 4.9
Onychoschizia.

also by climatic and seasonal factors, which influence the hydration of the nail.

Thinning of the nail plate can be the result of reduction in the length of the matrix by crescent biopsy, or due to disease, such as eczema, lichen planus, psoriasis, and in disorders of the peripheral circulation. Brittle nails are frequent in alopecia areata.

Systemic causes. These include hypochromic anaemia, arsenic poisoning, infectious diseases, severe toxaemia, rheumatoid arthritis, vitamin deficiency, particularly of A, C, and B6, osteoporosis, and osteomalacia. Finally, there are many inherited diseases associated with nail atrophy (Shelley and Rawnsley, 1965).

Transient nail fragility may be due to exposure to cold, emotional disturbances, and attacks of depression. The subsequent appearance of a Beau's line enables the retrospective diagnosis, but frequently the cause of these short-lived episodes remains obscure.

BIBLIOGRAPHY

[1] ACHTEN G., WANET-ROUARD J. — Atrophie unguéale et trachyonychie. *Arch. belg. Derm.*, 1974.
[2] ALKIEWICZ J. — Trachyonychie. *Ann. Derm. Syph. (Paris)*, 1950, *10*, 136–140.
[3] BARAN R. — Les ongles fragiles. *Les nouvelles esthétiques*, Sept. 1977, 41–45.
[4] BARAN R., DUPRÉ A.— Twenty nail dystrophy of childhood. *Arch. Derm. (Chicago)* 1977, *113*, 1613.
[5] BEAU J.-H.-S. — Certains caractères de séméiologie rétrospective présentés par les ongles. *Arch. Gén. Méd.*, 1846, *9*, 447.
[6] DUBREUILH W., FRECHE. — Onychorrhexis p. 845–853, in *3ᵉ Congrès international de Dermatologie*. Waterlow (London) Ltd, 1896.
[7] LECLERCQ R. — Naevus striatus symetricus unguis, dystrophie médiane canaliforme de Heller ou dystrophie unguéale médiane en chevrons. *Bull. Soc. fr. Derm.*, 1964, *71*, 654–658.
[8] SAMMAN P.-D.— *The nails in disease*. London, William Heinemann Med. Books, 1978, 3ᵉ edit.
[9] SHELLEY W.-B., RAWNSLEY H.-M. — Aminogenic alopecia. *Lancet*, 1965, *2*, 1327–1328.

5. *MODIFICATIONS OF COLOUR: CHROMONYCHIAS OR DYSCHROMIAS*

R. Baran

Generally, anomalies of colour depend on the transparency of the nail, its attachments and the character of the underlying tissues. Colour is also affected by the state of the skin vessels and the composition of the blood.

White nails are the most common of the variants, and can be divided into two main types – true leukonychia and apparent leukonychia.

In true leukonychia, the nail appears opaque and white in colour owing to diffraction of light on imperfect cornification of the remnants of the horny cells. It may be complete leukonychia totalis (rare) or incomplete (subtotal leukonychia) (Plate 1). Partial forms are divided into punctate leukonychia, which is common; striated leukonychia, relatively common; and distal leukonychia, which is very rare.

In apparent leukonychia, there are three types, depending on site of involvement:

(a) as a result of infection of the nail plate, for example, pseudo-leukonychia due to superficial white fungal infections
(b) as a result of onycholysis
(c) as a result of disease of the matrix and/or the nail bed: for example, pseudo-macro-lunula.

TRUE LEUKONYCHIA

TOTAL LEUKONYCHIA

The nail may be milky, chalky, bluish, ivory or porcelain white in colour. The lucency of the whiteness varies. When it is faintly opaque, it may be possible to see striated transverse leukonychia crossing a nail attacked by complete leukonychia. In polarised light, the nail structure appears disorganised, due to disorientation of the keratin fibrils (Baran and Achten, 1969).

SUBTOTAL LEUKONYCHIA

In this form, there is a pink arc of about 2–4 mm width proximal to the white area. This can be explained by the fact that the nucleated cells in the proximal area mature, lose their keratohyaline granules and then produce healthy keratin several weeks after they have been formed.

However, Juhlin (1963), believes that there are nucleated cells along the whole length of the nail, whose number decreases as they approach the distal end, thus producing the normal pink colour to the point of separation from the nail bed. There are, however, still enough left, so the theory goes, for the nail to acquire a whitish tint when it has lost contact with the nail bed.

Leukonychia, whether it be total or subtotal, may be permanent or temporary according to its origins.

STRIATED LEUKONYCHIA

Here one or several nails show a band, usually transverse, 1 or 2 mm wide and usually at the same site in each nail (Plate 2). There is also a rare longitudinal striated leukonychia, the result of a parakeratotic hyperplasia of the nail bed epidermis, with or without abnormal keratinisation of the deeper cells of the nail plate (Higashi *et al.*, 1971).

TABLE 5.1 Exogenous chromonychias

Aetiology	Colour	Location	Remarks
1. Occupational			
butchers	white	bed	parakeratosis of nail bed
coffee roasters	brown	nail	
hairdressers	brownish	nail	hair dyes
carpenters	brown	bed	wood (ebony)
vineyard workers	dark yellow	nail	cultivation
photographers (Plate 6)	brown	nail	
gunsmiths	brown	nail	
french polishers	brown	bed	
confectioners (caramels)	brown	nail	

(*continued on next page*)

TABLE 5.1 (*continued*)

Aetiology	Colour	Location	Remarks
dyers	brown	nail	⎰ Dinitroorthocresol Dinobuton
repeated handling or abnormally high concentration			⎱ Diquat Paraquat
agricultural chemicals	white or yellow	nail	

2. *Mixed (occupational and/or therapeutic)*

occupational radiotherapy[16]	transitory blueness, brown;	nail nail	Wohlbach's triad (hyperdyskeratosis, telangiectasia, sclero-atrophy), deformed,
	'coal spots'	bed (micro-haematomas)	thickened or thinned nail, longitudinal striations, onychorrhexis with splintering of the free edge, onychoschizia and recurring paronychia
local therapeutic radiotherapy[44,45]	brown	nail	transverse or longitudinal pigmented bands
general radiotherapy (antitumoural)[22] (antitumour agents)[22]	brown	nail	transient pigmentation in form of wide transverse band

3. *External applications*

silver nitrate	grey	nail	
mercury bichloride[12]	grey	nail	
ammoniated mercury[11]	brown	nail	
potassium permanganate	chestnut brown	nail	
chrysarobine, anthralin, resorchin	orange	nail	
mercurochrome, eosin	red	nail	
gentian violet	violet	nail	
fuschin	purple	nail	
methylene blue	blue	nail	
methyl green	green	nail	
fluoresceine	yellow	nail	
picric acid	yellow	nail	

4. *Other external factors*

cosmetics:			
nail varnish (Plate 8)[28]	orange-brown	nail	
henna	chestnut brown	nail	
smoking (nicotine)	chestnut brown	nail	along the lateral nail fold and adjacent pulp, first, second and third fingers

5. *Trauma*

manicure	partial leukonychia	nail	sometimes transverse furrow
cryotherapy	partial leukonychia	nail	
extensive haematoma	black	nail/bed	common in big toes due to
splinter haemorrhages	black	nail/bed	repeated minor injury, bilateral (sportsman's toe).

Superior numbers refer to references in the bibliography.

TABLE 5.2 Toxic or therapeutic endogenous chromonychias

Aetiology	Colour	Location	Remarks
Metallic salts:			
gold	dark brown	nail	
silver	slate blue	lunula and bed	
lead	partial leukonychia	nail	
bichromates	yellow ochre	nail	
Phenothiazines[41]	brown	bed	strong doses and prolonged treatment (psychiatric)
Antimalarial drugs:			
quinacrine (Plate 9)[25]	blue–black	bed	
	white, yellow, grey }	nail and bed	fluorescence of the nails in UV light
chloroquine[18]	purplish-blue	bed	
camoquine (Plate 10)	blue–grey	bed	
Tetracyclines (Plate 11)	yellow	nail	prolonged treatment, fluorescence in UV light
dimethylchloro-tetracycline[42]	brown	nail	importance of the triad: photosensitisation-dyschromia-onycholysis for the first two cyclines
doxycycline (Plate 12)	brown	nail	
chlor- and oxytet-racycline	brown	nail	
Sulphonamides	partial leukonychia	nail	
Phenolphthalein[13]	dark grey	lunula	
Antimitotics			
busulfan (Misulban)	brown	bed (proximal attack)	
cyclophosphamide	brown	nail	
5-fluoro-uracil	brown	nail	
hydroxyurea	melanic slate	nail and bed	brittle atrophied nails, onychoschizia
doxorubicin (adriamycin)	brown	bed	onycholysis
bleomycin	brownish striations	nail	onycholysis
Acetyl-acetic acid	purpura	bed	
Neosynephrine	purpura	bed	
Trinitrotoluene	purpura	bed	
Aniline			
and nitric derivatives of benzinic carbides	purplish-blue with cyanosis	bed	methaemoglobinaemia
Carbon monoxide	cherry red	bed	cochineal pink skin
Arsenic	brownish, striated	nail	complete or partial. Mees' bands,
	leukonychia	nail	classic but rare (acute intoxication)

(*continued on next page*)

TABLE 5.2 *(continued)*

Aetiology	Colour	Location	Remarks
Thallium	brownish striated leukonychia	nail	complete or partial Mees' band (acute intoxications)
Beta-carotene	yellow	bed	

Superior numbers refer to references in the bibliography.

PUNCTATE LEUKONYCHIA

This consists of spots of 1–3 mm in diameter which occur singly or in groups. This appearance is usually due to occupational or other (for example, manicure) microtrauma to the matrix. The evolution of the spots is variable, appearing generally on contact with the cuticle: they grow distally with the nail but about half disappear in the course of their migration towards the free edge. This proves that parakeratotic cells are capable of maturing and losing their keratohyaline granules to produce keratin, although they have not been vascularised for months. Other white spots enlarge and others appear at a distance from the lunula, suggesting that the nail bed is participating by incorporating groups of nucleated cells into the nail. A similar process could explain the exclusively distal leukonychias which have occasionally been observed (Juhlin, 1963).

A local or general fault in normal keratinisation is not the only cause of punctate leukonychia. Infiltration of air, as is known to occur in cutaneous parakeratoses, perhaps may play a part. When considering aetiology it is necessary to make a distinction between disorder of the nail itself and of the nail bed.

LEUKONYCHIA CAUSED BY NAIL PLATE INVOLVEMENT OR TRUE LEUKONYCHIA

Congenital forms. These are transmitted as autosomal dominants. They are usually total or subtotal and rarely punctate or striated. These congenital forms can be associated with other malformations of the nail, skin or other tissues (e.g. deafness).

ACQUIRED FORMS

(a) *Of external origin.* Over-zealous manicuring is the main cause of this punctate leukonychia, so common in women. This can also produce transverse white striae (Plate 8). There are also occupational causes (Plate 6): complete leukonychia or longitudinal striae occurs in workers in pickling factories.

(b) *Of endogenous origin.* Endogenous leukonychia occurs:
  (i) after acute diseases, such as infectious diseases (measles), cardiac diseases (myocardial infarction), diseases of the alimentary tract (ulcerative colitis), renal diseases, and also serious shock, fractures, and surgery
  (ii) during chronic diseases, such as renal insufficiency, where Hudson and Dennis have observed that transverse bands indicate the severity of the illness

  (iii) following poisoning (Plates 7 to 12) by thallium, arsenic, lead, the sulphonamides and pilocarpine. In acute arsenic poisoning, Mees' bands (Plate 2), small transverse white lines occurring at the same site in each nail, are of medico-legal interest; in chronic arsenic poisoning, white diagonal striae are said to be equally characteristic, but these classical opinions are not held by Ronchèse (1951), with whom we agree
  (iv) in physiological phenomena, like menstruation (striated leukonychia) (Plate 3).

APPARENT LEUKONYCHIA, DUE TO DISEASE OF THE SUBUNGUAL TISSUES

TERRY'S WHITE NAIL

Terry (1954) was the first to describe an opacity of the nails in 82 out of 100 patients with cirrhosis (Plate 4). In the majority of cases, the nails are of an opaque white colour, obscuring the lunula. This colouring, which stops suddenly 1–2 mm from the distal edge of the nail, leaves a pink area corresponding to the onychodermal band. It lies parallel to the distal part of the nail bed and may be irregular. Terry's nails involve all nails uniformly.

MOREY AND BURKE'S NAIL (1955)

This is a variation of Terry's nail. The authors report four cases in which the whitening of the nail extended to the middle part with a curved frontal edge. One of the cases had identical changes in the toes.

MUEHRCKE'S PAIRED NARROW WHITE BANDS (1956)

These bands, which are parallel to the lunula, are separated from one another and from the lunula by strips of pink nail. These paired narrow white bands disappear when the serum albumin level returns to normal and reappear if it falls again.

It is possible that hypoalbuminaemia produces oedema of connective tissue in front of the lunula just below the epidermis of the nail bed, changing the compact arrangement of the collagen in the area into a looser texture resembling the structure of the lunula, and hence the whitish colour. The correlation between the presence or disappearance of the white bands, and the amount of serum albumin, seems to confirm this hypothesis. Unfortunately, there is no information about the relationship between the serum albumin level and white cirrhotic nails, except that supplied by Morey and Burke about their four patients, whose levels varied around 2 g per 100 ml.

URAEMIC HALF AND HALF NAIL

This nail, described by Lindsay (1957), consists of two parts separated more or less transversely by a well-defined line (Plate 5): the proximal area is dull whitish, resembling ground glass and obscuring the lunula; the distal area is pink, reddish or brownish, and occupies between 20–60 per cent of the total length of the nail (average 33 per cent). In typical cases diagnosis presents no difficulty, but in Terry's nail the pink distal area may occupy up to 50 per cent of the length of the nail, in which case the two types of nail may be confused.

DERMATOLOGICAL FORMS OF LEUKONYCHIA

In psoriasis the nail may be affected by true leukonychia, due to involvement of the matrix, and apparent leukonychia, due to onycholysis, and parakeratotic deposits in the nail bed. One of the earliest signs of leprosy is a pseudo macrolunula (Pardo-Castello, 1960). In dystrophic leprosy, this apparent leukonychia may become total. Leukonychia may also occur in other dermatoses, such as alopecia areata, dysidrosis, Darier's disease.

TABLE 5.3 Dermatological chromonychias

Aetiology	Colour	Location	Remarks
Longitudinal Melanotic streaks[20,26]	brownish	matrix and nail	sometimes existence of a junctional naevus or active melanocytes
Peutz-Jeghers-Touraine syndrome[49]	black	matrix and nail	longitudinal bands (active melanocytes), possible clubbing
Malignant melanoma	black	matrix and nail	existence of post-traumatic forms with initial haematoma
Glomus tumour	dark blue	bed	painful syndrome
Angioma – Cot's syndrome	bluish or reddish	bed	
L.E.O.P.A.R.D. syndrome[43]	leukonychia	nail	leukokoilonychia
Alopecia areata	partial leukonychia	nail	
Dysidrosis	partial leukonychia	nail	
Acanthosis nigricans	dull greyish	nail	brittle nails
Pachyonychia congenita	yellowish brown–yellow	nail	very hard pachyonychia with hypertrophy of the nail bed
Darier's disease	white and red longitudinal bands	nail bed	pathognomonic triad: Distal subungual keratoses and white and red bands
Pseudomonas[6,14,17,31]	green	nail	the pyocyanin (dark blue) is soluble in water and chloroform; the fluorescein (greenish-yellow) is only soluble in water. Blue–green fluorescence in UV light.

(continued on next page)

TABLE 5.3 (*continued*)

Aetiology	Colour	Location	Remarks
Proteus mirabilis	black	nail	
Acrothecium nigrum[50]	black	bed	
Alternaria grisea tenius	black	nail	
Hormodendrum elatum	black	nail	
Blastomycetes	green–brown	nail	
Scopulariopsis	yellowish	nail	
Candida[17]	yellowish-white, brown, green (?)	nail	even in the absence of *Pseudomonas*, test to see if the water is coloured green
Trichophyton mentagrophytes (interdigitale)	leukonychia black	bed, nail	
Trichophyton mentagrophytes (interdigitale)	leukonychia	nail	superficial white onychomycosis[51]
Cephalosporium	leukonychia	nail	superficial white onychomycosis[51]
Aspergillus terreus	leukonychia	nail	superficial white onychomycosis[51]
Aspergillus flavus, nidulans, etc.[17]	greenish (?)	nail	

Superior numbers refer to references in the bibliography.

TABLE 5.4 Endogenous chromonychia, general non-infectious diseases

Aetiology	Colour	Location	Remarks
1. Digestive origin			
Laënnec's disease	white	bed	Terry's 'white nail'[46]
Wilson's disease	bluish	lunula	
Jaundice	yellow	bed	
Hyperbilirubinemia (Aplas' type)	dark brownish	nail	
Haemochromatosis (Plate 13)	total leukonychia	nail	with koilonychia (50% of cases)[10]
	pseudo-macrolunula	bed	striations, fragility
	grey nails	nail	rare, bilateral clubbing,
	brown nails	nail	hyperpigmentation often starts on the proximal nail fold with a thin brown periungual ring[24]

(*continued on next page*)

TABLE 5.4 (*continued*)

Aetiology	Colour	Location	Remarks
2. Cardio-vascular diseases			
Venous stasis	cyanotic	bed	
Cardiac insufficiency[47]	cyanotic red	bed lunula	Terry's 'red lunula'
Arterial ischaemia (gangrene)	black	bed	
Samman and White's syndrome[40]	yellow	nail	lymphatic anomalies
3. Modifications to the composition of the blood			
Anaemia	pale pink	bed	koilonychia in hypochromic anaemia
Polycythemia	dark red	bed	
Pulmonary insufficiency	cyanotic	bed	
Hypoalbuminaemia[33]	white	bed	Muehrcke's paired narrow white bands
Methaemoglobinaemia	cyanotic	bed	
Cryoglobulinaemia (Plate 14)	purpura	bed	
4. Metabolic diseases			
Porphyria cutanea tarda[15,35]	yellow brownish blackish	nail and/or bed	usually distal, subungual serosanguinous bullae, absence of lunula, early koilonychia
Erythropoietic protoporphyria[36,48] }	opaque, grey–blue brown	nail	possible absence of lunula
Congenital porphyria	brownish	nail	
Ochronosis	grey–blue	bed	appears in adulthood
Acute renal insufficiency	partial leukonychia	nail	
Chronic renal insufficiency:			
azotemic onychopathia, Lindsay's type[27]	pseudo-macrolunula, pink distal area, red or brown	bed	in the distal arc: either increase in the number of capillaries and thickening of their walls; or melanoric granules in the basal layer of the nail bed
azotemic onychopathia, Leyden's type	brown distal area	nail	pigment throughout the distal part of the nail
5. Others			
Menstrual periods			
Surgery, fractures, ulcerative colitis, myocardial infarctions	partial leukonychia	nail	
Trophic problems	grey or blackish	nail	
Nervous injury	brown	nail and bed	
Deficiency of B12[5]	brown	nail	
Malnutrition[8]	pigmented bands	nail	
Addison's disease[1,9]	pale nails brownish sometimes	bed nail	due to anaemia either longitudinal bands or diffuse pigmentation (caused by MSH or ACTH), nail fragility (demineralisation)

Superior numbers refer to references in the bibliography.

TABLE 5.5 Infectious and parasitic endogenous chromonychias

Aetiology	Colour	Location	Remarks
Syphilis	amber	nail	as rare as it is classic
	brownish	nail	
	distal lilac arc	between nail bed and hyponychium	may correspond to 'onychodermal band', varies in colour and width
Pinta[29]	brown	bed	
Leprosy	white	bed	spreading of lunula[34]
Subacute bacterial endocarditis	splinter haemorrhages	ungual and subungual haemorrhages	microbial embolism
Trichinosis	splinter haemorrhages	ungual and subungual	suddenly appears on several fingers
Malaria	slate grey	bed	immediately precedes and accompanies febrile crisis
Viral or spiro-chaetal jaundice	yellow	bed	
Acute infections	partial leukonychia	nail	

Superior numbers refer to references in the bibliography.

BIBLIOGRAPHY

[1] ALLENBY C.-F., SNELL P.-H. — Longitudinal pigmentation of the nails in Addison's disease. *Brit. med. J.*, 1966, *1*, 1582–1583.

[2] BARAN R., ACHTEN G. — Les associations koïlonychie et de leuconychie totale. *Arch. belges Derm.*, 1969, *25*, 13–29.

[3] BARAN R. — Onychopathie de contact aux produits phyto-sanitaires organo-synthétiques. *Bull. Soc. franç. Derm. Syph.*, 1973, *80*, 172–173.

[4] BARAN R., GIOANNI T. — Half and half nail (Ongle équisegmenté azotémique). *Bull. Soc. franç. Derm. Syph.*, 1968, *75*, 399–400.

[5] BAKER S.-J., IGNATIUS M., JOHNSON S. — Hyperpigmentation of skin: A sign of vit. B12 deficiency. *Brit. med. J.*, 1963, *1*, 1713–1715.

[6] BAUER M.-F., COHEN B.-A. — The role of pseudomonas aeruginosa in infection about the nails. *Arch. Derm. (Chic.)*, 1957, *75*, 394–396.

[7] BEARN A.-G., McKUSICK V.-A. — Azure lunulae-an usual change in the fingernails in two patients with hepatolenticular degeneration (Wilson's disease), *JAMA*, 1958, *166*, 903–906.

[8] BISHT D.-B., LUCKNOW M.-D., SINGH S.-S., RAJASTHAN M.-B. — Pigmented bands on nails a new sign in malnutrition. *Lancet*, 1962, *1*, 507–508.

[9] BISSEL G.-W., SURAKOMOL K., GREENSLIT F. — Longitudinal banded pigmentation of nails in primary adrenal insufficiency. *JAMA*, 1971, *215*, 1666–1667.

[10] BOUREL M., SIMON M., PAWLOTSKY Y., MURIE N., LE CARRERES D. — Altérations unguéales dans l'hémochromatose idiopathique. *Sem. Hôp. Paris*, 1970, *46*, 677–680.

[11] BUTTERWORTH T., STREAN L.-P. — Mercurial pigmentation of nails. *Arch. Derm.*, 1963, *88*, 55–57.

[12] CALLAWAY J.-L. — Transient discoloration of the nails due to mercury bichloride. *Arch. Derm.*, 1937, *36*, 62–64.

[13] CAMPBELL G.-S. — Peculiar pigmentation following the use of a purgative containing phenolphthalein. *Brit. J. Derm.*, 1931, *43*, 186–187.

[14] CHERNOSKY M.-E., DUKES C.-D. — Green nails. *Arch Derm.*, 1963, *88*, 548–553.

[15] DURET R.-L., CORNIL A., THYS O. — A propos de 36 cas de porphyropathie. *Acta. clin Belg.*, 1968, *23*, 114.

[16] EPSTEIN E. — *Radiodermatitis*. Springfield (III), Charles C. Thomas, 1962.

[17] GOLDMAN L., FOX H. — Greenish pigmentation of nail plates from bacillus Pyocyaneous infection. *Arch. Derm. Syph. (Chic)*, 1944, *49*, 136–137.

[18] GRACIANSKY P., GRUPPER C. — Pigmentation anormale cutanéo-muqueuse et sous-unguéale au cours du traitement par la nivaquine et la flavoquine de deux cas de lupus érythémateux chronique. *Bull. Soc. franç. Derm. Syph.*, 1956, *5*, 444.

[19] HEARN C.-E.-D., KEIR W. — Nail damage in spray operators exposed to Paraquat. *Brit. J. industr. Med.*, 1971, *28*, 399.

[20] HIGASHI N. — Melanocytes of nail matrix and nail pigmentation. *Arch. Derm.*, 1968, *95*, 570–574.

[21] HIGASHI N., SUGAI T., YAMAMOTO T. — Leukonychia striatus longitudinalis. *Arch. Derm. (Chic.)*, 1971. *104*, 192–196.

[22] INALSINGH A. — Melanonychia after treatment of malignant disease with radiation and cyclophosphamide. *Arch. Derm. (Chic.)*, 1972, *106*, 765.

[23] JUHLIN L. — Hereditary leukonychia. *Acta derm. venereol. (Stockh.)*, 1963, *43*, 136–141.

[24] KALK H.-O. — Uber Hautzeichen bei Leberkrankherten. *Dtsch. med. Wschr.*, 1957, *38*, 1637–1641.

[25] KIERLAND R.-R., SHEARD C., MASON H.-L., LOBITZ W.-C. — Fluorescence of nails from quinacrine hydrochloride. *JAMA*, 1946, *131*, 809–810.

[26] LEYDEN J.-J., SPOTT D.-A., GOLDSCHMIDT H. — Diffuse and banded melanin pigmentation in nails. *Arch. Derm.*, 1972, *105*, 548–550.

[27] LINDSAY P.-J. — The half-and-half nail. *Arch. intern. Med.*, 1967, *119*, 583–587.

[28] LOVEMAN A.-B., FLIEGELMAN M.-T. — Discoloration of nails. *Arch. Derm.*, 1955, *72*, 153–156.

[29] MEDINA R. — El carate en Venezuela. *Derm. Venezolana*, 1963, *3*, 160–230.

[30] MITCHELL J.-C. — A clinical study of leukonychia. *Brit. J. Derm.*, 1953, *65*, 121–130.

[31] MOORE M., MARCUS M.-D. — Green nails: Role of Candida (Syringospore, Monilia) and Pseudomonas aeruginosa. *Arch. Derm. Syph.* (*Chic.*), 1951, *64*, 499–505.

[32] MOREY D., BURKE J. — Distinctive nails changes in advanced hepatic cirrhosis. *Gastroenterology*, 1955, *29*, 258–261.

[33] MUEHRCKE R. — The finger-nails in chronic hypoalbuminaemia. *Brit. Med. J.*, 1956, *1*, 1327–1328.

[34] PARDO-CASTELLO V., PARDO O. — *Disease of nails.* Springfeld (3 edit.), Charles C. Thomas, 1960.

[35] PUISSANT A., DAVID V., LACHIVER D., AITKEN G. — Formes cliniques atypiques de la porphyrie cutanée tardive. *Boll. Instituto Derm. S. Gallicano*, 1971, 7, 19–30.

[36] REDEKER A.-G., BERKE M. — Erythropoietic protoporphyria with eczema solare. *Arch. Derm.*, 1962, *86*, 569.

[37] RITCHIE E.-B., PINKERTON M.-E. — Fusarium oxysporum infection of nail. *Arch. Derm.* (*Chic.*), 1959, *79*, 705–708.

[38] RONCHESE F. — Pecular nail anomalies. *Arch. Derm.* (*Chic.*), 1951, *63*, 565–580.

[39] SAMMAN P., JOHNSTON E.-N.-M. — Nail damage associated with handling of paraquat and diquat. *Brit. Med. J.*, 1969, *1*, 818–819.

[40] SAMMAN P., WHITE W. — The yellow nail syndrome. *Brit. J. Derm.*, 1964, *76*, 153–157.

[41] SATANOVE A. — Pigmentation due to phenothiazines. *JAMA*, 1965, *191*, 263–268.

[42] SEGAL B.-M. — Photosensitivity nail discoloration and onycholysis. *Arch. Intern. Med.*, 1963, *112*, 165.

[43] SELMANOVITZ V.-J., ORENTREICH N. — Lentiginosis profusa in daughter and mother: Multiple granular cell «myoblastomas» in the former. *Arch. Derm.* (*Chic.*), 1970, *101*, 615–616.

[44] SHELLEY W.-B., RAWNSLEY H.-M., PILLSBURY D.-M. — Postirradiation melanonychia. *Arch. Derm.*, 1964, *90*, 174–176.

[45] SUTTON R.-L. — Transverse band pigmentation of fingernails after X-ray therapy. *JAMA*, 1962, *150*, 210.

[46] TERRY R. — White nails in hepatic cirrhosis. *Lancet*, 1954, *1*, 757–759.

[47] TERRY R. — Red half-moons in cardiac failure. *Lancet*, 1954, *2*, 842–844.

[48] THIVOLET J., FREYCON J., PERROT H., GUIBAUD P., BEYVIN A.-J. — Protoporphyrie érythropoïétique. *Bull. Soc. franç. Derm. Syph.*, 1968, *75*, 829–841.

[49] VALERO A., SHERF K. — Pigmented nails in Peutz-Jeghers syndrome. *Amer. J. Gastroent.*, 1965, *43*, 56–58.

[50] YOUNG W.-J. — Pigmented mycotic growth beneath the nail. *Arch. Derm.* (*Chic.*), 1934, *30*, 186–189.

[51] ZAIAS N. — Superficial white onychomycosis. *Sabouraudia*, 1966, *5*, 99–103.

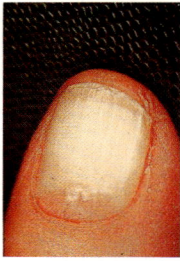

1

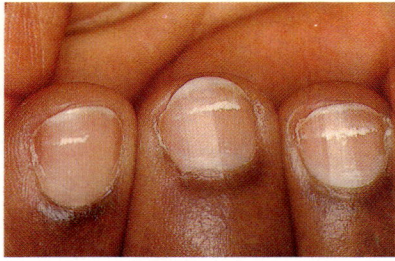

2

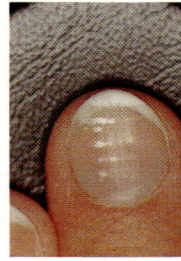

3

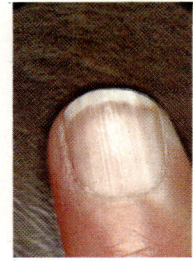

4

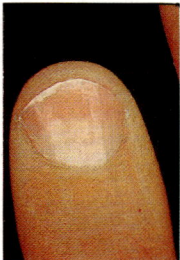

5

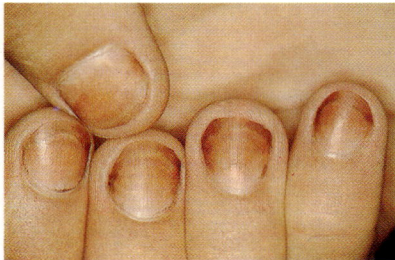

6

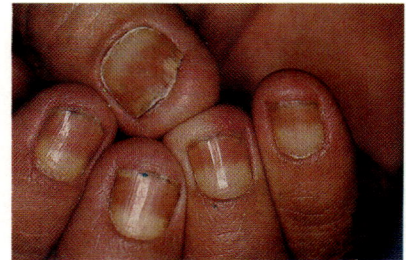

7

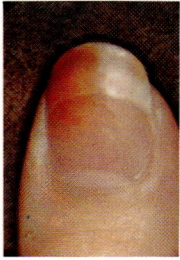

8

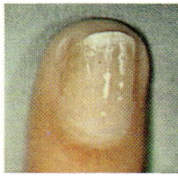

9

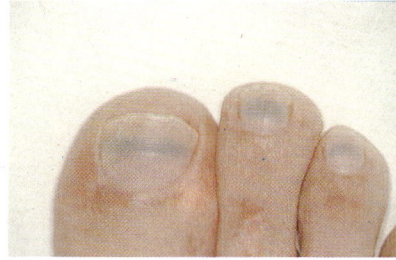

10

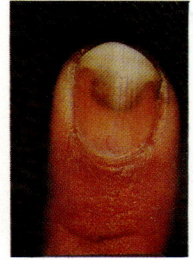

11

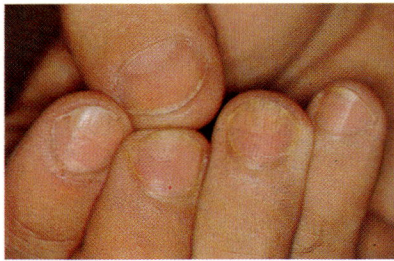

12

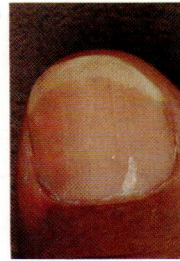

13

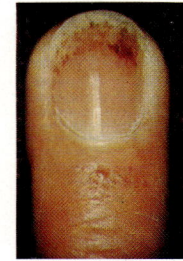

14

Plate 1 Superlunary and subtotal leuconychia

Plate 2 Striatal leuconychia of unknown aetiology calling to mind Mees' classic arsenical bands

Plate 3 Menstrual striatal leuconychia

Plate 4 Terry's white cirrhotic nail which respects a pink distal arch

Plate 5 Hyperazotemic equisegmented nail (Baran) or Lindsay's half-and-half nail

Plate 6 Professional chromonychia (photographer)

Plate 7 Onychopathy by organosynthetic phytosanitary products (dinofro-ortho-cresol)

Plate 8 Chronomonychia due to cosmetics (nail varnish)

Plate 9 Pseudo-chronomonychia due to drugs (quinacrine) (Laugier's collection)

Plate 10 Pseudo-chronomonychia due to antimalarial drugs (flaroquine). The nail bed is blue-green

Plate 11 Photo-onycholysis due to tetracycline (rare)

Plate 12 Photo-onycholysis due to doxicycline (common). The attack can be seen particularly clearly on the 2nd, 3rd and 4th fingers

Plate 13 Haemochromatosis. Brown nail and pigmented skin in a white person

Plate 14 Cryoglobulinemia, purpura and filiform haematomae (Schnitzler's collection)

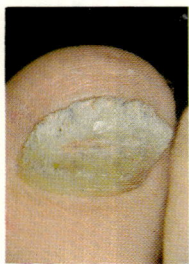 15 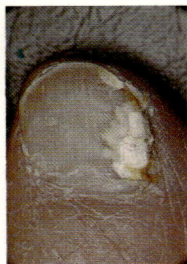 16  17 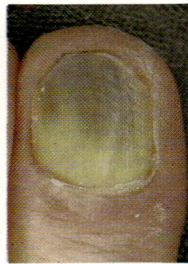 18 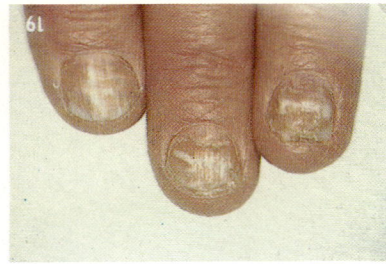 19

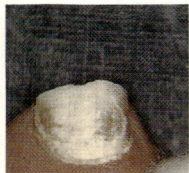 20 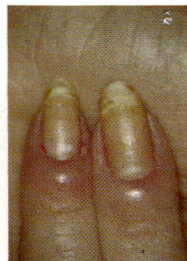 21 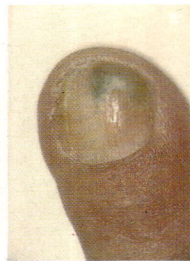 22 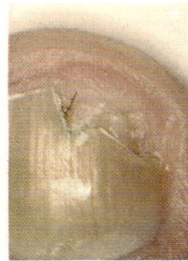 23 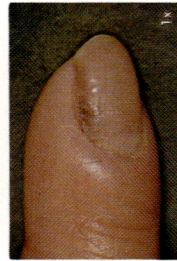 24 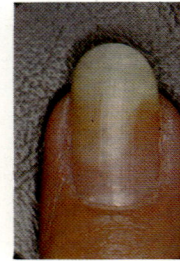 25

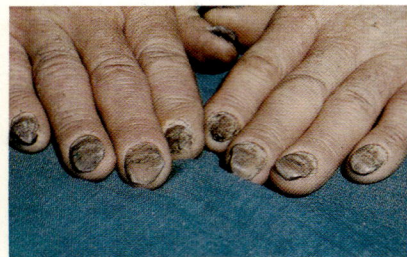 26 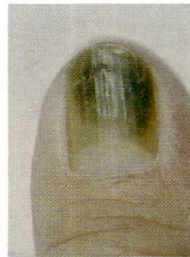 27  28 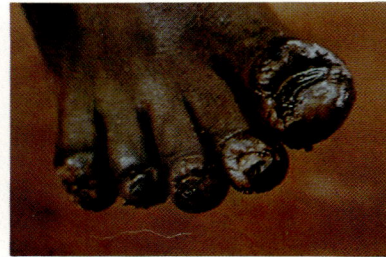 29

Plate 15 Trichophyton rubrum (toes)

Plate 16 Trichophyton rubrum and Candida albicans (toes)

Plate 17 Interdigital trichophyton (toes)

Plate 18 Scopularopsis and Candida albicans (toes)

Plate 19 Trichophyton rubrum (fingers)

Plate 20 Superficial leuconychomycosis with Aspergillus

Plate 21 Perionyxis with Candida albicans

Plate 22 Onycholysis caused by foreign body (hair) Aspergillus niger

Plate 23 Aspergillus niger (same patient)

Plate 24 Paronychic candidiasis with secondary onychia

Plate 25 Pure candidiasic onycholysis

Plate 26 Primary dystrophic onychia (Agache's collection)

Plate 27 Pseudomonas

Plate 28 Solubility test for pyocyaneus pigments

Plate 29 Parasitic perionyxis caused by female chigoe

6. ONYCHIA AND PARONYCHIA OF MYCOTIC, MICROBIAL AND PARASITIC ORIGIN

R. Baran

TABLE 6.1 Mycotic onychias

Principal clinical types	Mycobacteriological examinations	Frequency	Location	Appearance
I Distal subungual onychomycosis	Dermatophytes: *Trichophyton rubrum*, especially *Trichophyton mentagrophytes* (interdigitale)	25%	hands and feet	Sometimes occurs as type of onycholysis, or else in form of subungual keratosis, pushing up free edge of nail under which there is a varied microbial flora. The disease spreads under the nail towards proximal area and nail loses its transparency. Its colour changes from chestnut brown to green or grey. Nail becomes friable like rotten wood. In final stage it crumbles or disappears after an injury. The nail bed, which is very thickened and abnormal, may still have fragments of nail: this stage is total secondary dystrophic onychomycosis (type IV).
	Moulds: *Scopulariopsis cephalosporium*, *Aspergillus fusarium*		feet	
	Yeasts: *Candida albicans*, *parapsilosis* (USA), *Candida tropicalis* (France)	50%	hands and feet	
	Role played by some bacteria? (N. Zaias, 1969, 1972)	25% (USA) very common	feet	
II Superficial white onychomycosis	Dermatophytes: *Trichophyton mentagrophytes* (interdigitale)	++++		Little white opacities with distinct edges coalesce and gradually cover whole nail surface. Latter is very rough, softer than usual and crumbles easily. Old lesions turn yellow.
	Moulds: *Cephalosporium*, *Aspergillus fusarium*	± rare	feet	
III Proximal white subungual onychomycosis	Dermatophytes: *Trichophyton rubrum*, *megnini, schoenleini, tonsurans, mentagrophytes* (interdigitale)	± very rare	hands or feet	Lesions are found beneath the nail. They appear under subungual fold in form of white areas, confined initially to lunula area and extend distally to involve whole nail.
IV Total secondary dystrophic onychomycosis	Dermatophytes Yeasts: *Candida albicans*	fairly frequent	hands ++ or feet hands and feet	Above all advanced form of type I.

(continued on next page)

TABLE 6.1 (*continued*)

Principal clinical types	Mycobacteriological examinations	Frequency	Location	Appearance
V Total dystrophic primary onychomycosis	*Candida albicans* Chronic mucocutaneous candidiasis	exceptional	all fingers and toes	Sometimes nail is brownish, usually thick and opaque, non-friable and striped lengthwise with whitish bands. Marked dermal inflammatory reaction with thickening of nail bed and hyponychium explaining pseudo-clubbing. All the nail structures are invaded by *Candida*. Possibility of paronychia; frequent and extensive mucocutaneous involvement.
Candida granuloma	Acrodermatitis enteropathica Hypoparathyroidism Hypothyroidism Addison's disease Ovarian insufficiency Thymoma			

ONYCHOMYCOSES

Table 6.1 is based on the modified classification of mycoses by Zaias (1966, 1969, 1972), according to the site of fungal attack (Figs. 6.1 and 6.2). In type I (Plates 15–19), that is distal subungual onychomycosis, the initial invasion occurs in the epithelial layer between the hyponychium and the nail bed. In type II (Plate 20), superficial white onychomycosis, the nail plate is attacked. In type III, proximal white subungual onychomycosis, the proximal fold is invaded.

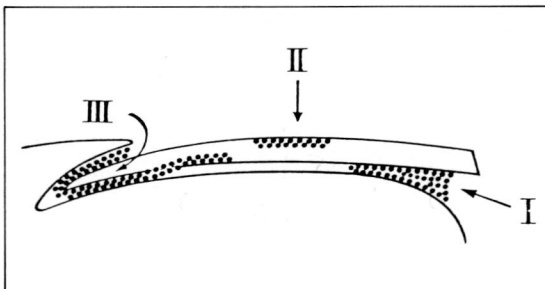

Figure 6.1
Areas attacked by fungi.
I. Distal subungual onychomycosis.
II. Superficial white onychomycosis.
III. Proximal white subungual onychomycosis.

Although it is impossible to tell the species of fungus from the appearance of the lesion, this does provide clues which allow the elimination of some genera and even some species. These observations give important clinical information, but precise diagnosis depends on laboratory identification of the fungus.

As with all mycological examinations, the examination of a nail consists of direct microscopic observation and culture. Scraping must be postponed for several days or weeks if local antifungal treatment has been applied. The tissue must be taken from the region where it is believed that the active mycelial filaments or

yeasts are in contact with healthy tissue for dermatophytes and from the paronychium for *Candida*. Nail and subungual debris are obtained using a scalpel blade or a curette. The specimens are collected in a sterile Petri dish. It may be necessary to use strong scissors to obtain an awkward superficial nail fragment. The nails from the hands and the feet and skin biopsies must be collected separately. The disintegration of nail clippings in a special mill or the use of a dental drill can raise the success rate of cultures.

For direct microscopic examination, fragments of nail are teased out as small as possible in a drop of 20 or 30 per cent potassium hydroxide. After gentle heating in a bunsen flame the preparation is studied with a microscope. In the mycoses either mycelial filaments or yeasts may be seen. It is not always easy to distinguish between mycelial filaments of dermatophytes and pseudo-filaments of yeasts. *Scopulariopsis* forms filaments as well as spores of which the size and morphology permit immediate identification, but often it is only lodging in lesions resulting from attack by *Trichophyton rubrum* or *Trichophyton interdigitale* (in the toe-nails). By repeated scrapings on the same patient, the *Trichophyton* may be isolated eventually as the responsible agent: genuine onychia caused by *Scopulariopsis* is rare (it has, however, been shown experimentally that *Scopulariopsis brevicaulis* is capable of invading the nail *in vitro*).

Cultures done systematically on Sabouraud's medium and incubated at 28–30°C, or simply left on a bench at room temperature, are of great value. Yeasts grow in 48 hours. Their identification takes several hours or days. They are usually yeasts of the genus *Candida*, most commonly *Candida albicans*. The number of isolated colonies of yeast in each tube, which represents fairly accurately the severity of the infection, should be carefully counted. The dermatophytes grow much more slowly, taking from 1 to 3 weeks. Their identification can then take several more weeks (Tables 6.2, 6.3 and 6.4).

Scopulariopsis brevicaulis has been studied as an example of an occasional pathogen (Liautaud *et al.*, 1971). Sometimes 'contaminating' or 'saprophytic' fungi, such as *Penicillium* and *Aspergillus*, can be isolated from nails. These only grow

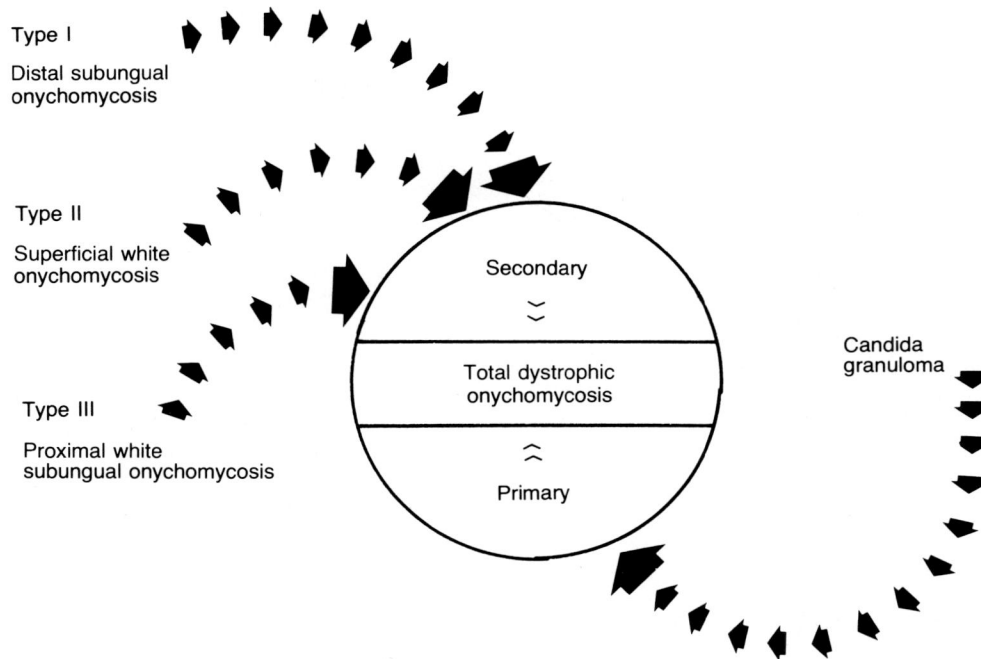

Type I

Distal subungual onychomycosis

Type II

Superficial white onychomycosis

Type III

Proximal white subungual onychomycosis

Secondary

Total dystrophic onychomycosis

Primary

Candida granuloma

Figure 6.2
Onychomycoses. Total dystrophic onychomycosis is the last stage of the onychomycoses, which develop right away in *Candida granuloma* infection, but gradually in the other types.

TABLE 6.2 Incidence of onychomycoses over three years (from G. Badillet, 1966) Hôpital Saint Louis pr P'Puïssant

| | Hands | | Feet | |
	Number	Percentage	Number	Percentage
Candida	498	79·6	126	12·8
Dermatophytes	121	19·3	684	69·4
Various fungi	7	1·1	175	17·8
Total	626		985	

TABLE 6.3 Incidence of onychomycoses due to dermatophytes over ten years (from G. Badillet) Hôpital Saint Louis pr P'Puïssant

Types	Hands	Feet	Total
T. rubrum	237	1075	1312
T. interdigitalie	7	299	306
E. floccosum	1	10	11
T. tonsurans	0	2	2
T. violaceum	3	0	3
T. soudanensis	1	0	1
T. sohoenleini	2	0	2
Total	251	1386	1637

TABLE 6.4 Incidence of onychomycoses due to candida over three years (from G. Badillet) (Hôpital Saint Louis pr P'Puïssant)

	Hands		Feet	
	Number	Percentage	Number	Percentage
Candida albicans	382	76·7	26	20
Candida tropicalis	89	17·9	80	63·5
Other Candida	27	5·4	20	15·9
Total	498		126	

on tissues where there is a primary lesion due to trauma for example (this explains the involvement of the big toe in half the cases), or due to disease of the nail (psoriasis, etc.); hence the name 'opportunists'. They cannot generally be considered pathogenic unless:

the filaments (usually deformed) can be seen by direct examination

the fungus has been isolated alone from all the infected areas without dermatophytes

the fungus is found at each scraping (Badillet, 1964, 1966) (at least 5/20 inocula must have yielded the mould).

Mycological examination of the recent onychia is easy. On the other hand, examination of an onychia which has developed over several years and has been treated with various antifungal agents is very difficult and sometimes unsuccessful.

To reduce the frequency of failure, Achten (1964) has proposed a technique of histological examination of the nail, by which a view of the whole thickness of the nail is obtained. Sometimes tunnels penetrating the keratin by the dermatophyte, corresponding to the transverse network of Alkiewicz, can be seen with the aid of a lens after treating the edge of the nail with cedar oil to make it clear. Where the network is dense it appears as an opaque white spot.

DIFFERENTIAL DIAGNOSIS

Psoriasis is the most important differential diagnosis. Generally it attacks all nails simultaneously, in contrast to dermatophytic onychomycoses in which at least one digit, or even a whole hand, is often spared.

Diagnosis is easy when the pathology is found in the classic sites of the disease. Diagnosis is also relatively easy when the dorsal aspect of the proximal fold is attacked but, if this is uninvolved and nail pits are absent, error is possible. Thus a specialised laboratory and correct interpretation of the information it provides are essential.

If cultures and histological examinations are negative, then psoriasis is probable. The recently developed study of the exfoliative cytology by Leyden et al. (1972) shows that in psoriasis there is a significant number of parakeratotic cells compared with the scarcity of nucleated cells in dermatophytic onychomycosis.

The keratin debris beneath the nail forms a favourable medium for the growth of micro-organisms. Candida parapsilosis is

prominent here. These are followed by contaminating fungi and by Gram positive and Gram negative (Pseudomonas) bacteria. However, dermatophytes are never found in psoriasis. This may be due to the inhibitory action of a specific glycoprotein on the latter organisms and not on the yeasts (Zaias), or perhaps due to the requirement of dermatophytes for normal keratin and their inability to colonise the parakeatotic skin (Grupper and Avram, 1972).

PARASITIC ONYCHIA: NORWEGIAN SCABIES

This disease is distinctive in that when it is selectively palmo-plantar and ungual, the degree of itchiness is inversely related to the intensity of the desquamation. Elbows, face and scalp are equally affected. There is a psoriasis-like subungual hyperkeratosis which is rich in mites.

BACTERIAL ONYCHIA

The development of a subungual infection usually as a result of trauma, whether or not associated with a paronychial infection, reduces the adherence of the nail plate and causes onycholysis. The various modifications of the nail plate which result from paronychial inflammation in the region of the matrix are described later.

PARONYCHIA

This is inflammation of the periungual folds. The acute form, of streptococcal or more often staphylococcal origin, is normally the result of trauma or a bad 'hangnail'. It results in a hot inflamed swelling which is sensitive to pressure. Delicate expression of the region produces a droplet of pus at the level of the lateral or posterior folds. The pain persists until the inflammatory lesion has healed as a result of medical treatment or surgical evacuation of the pus. It is not uncommon for the nail to fall off, starting at the matrix. Excess granulation tissue on the lateral fold commonly develops in relation to infection of an ingrowing nail.

Chronic paronychia starts as a red oedema in a lateral fold that is slightly painful. After several months or even years, the folds resemble a semi-circular cushion around the base of the nail which separates from the nail and retracts (Plate 21).

As a result of repeated episodes of paronychia, the inflammation of the matrix causes disturbance of the growth of the nail with changes in the appearance of the nail plate. The common effects are roughness, friability and/or transverse grooves and unevenness. This is sometimes associated with pigmentation of the nail.

After the paronychia has disappeared, secondary onychia may still develop. This normally regresses eventually, but sometimes it continues and, it may be difficult to distinguish this from primary onychia due to pyogenic bacteria or *Candida*.

Prolonged soaking is a factor in paronychia, hence the higher incidence of this disorder in women due to repeated contact with water, soap and detergent. Other aetiological factors are cold hands, manicuring, vaginal thrush and diabetes. This disorder is particularly common in some occupations: dish-washers, barmen (who also handle oranges and lemons), confectioners, fishmongers, etc.

Stone and Mullins (1964, 1968) consider that foreign material (debris of *Candida* organisms and foreign bodies associated with the occupation of the patient) contribute to the chronic nature of the lesions by invasion of the epidermis on the deep surface of the proximal nail fold. This explains the paronychia and distal dactylitis ('drumstick' hands) of chemical origin.

In children, the common predisposing factor is sucking of the thumb or fingers. Nevertheless, it is necessary to rule out certain diseases in which paronychia is commonly found: hypoparathyroidism, enteropathic acrodermatitis, coeliac disease and chronic cutaneo-mucous candidiasis with or without an immune deficiency.

The question of aetiology is not finally resolved. In agreement with Barlow *et al.* (1970) it is reasonable to allow that chronic paronychia is usually a mixed infection of *Candida albicans* and intestinal bacteria (*Streptococcus fecalis*, coliforms, *Proteus*, *Pseudomonas*). The various factors which damage the area allow pyogenic *Staphylococci* and *Candida albicans* to attack the keratin and cause the detachment of the cuticle (soft keratin) from the nail plate (hard keratin), but the frequency with which pyogenic *Staphylococci* are found indicates that they may be commensal. As a result, it is practically impossible to differentiate between chronic paronychia of pyogenic origin and of candidial origin, in the absence of cutaneo-mucous lesions (indicating fungal origin) and of discolouration of the nail plate.

It is interesting to evaluate the diagnostic importance of chromonychia.

The green nail syndrome, according to Goldman and Fox (1944), is the result of paronychial infection associated with *Pseudomonas* on the skin. Moore and Marcus (1951) include some fungi, such as various *Aspergillus* and *Candida* species, as causal agents.

1. Aspergillus

Although Moore and Weiss (1948) estimate that in the USA there are 12 species of *Aspergillus* capable of producing green colouration of the nail, in France there has only been a single instance in which a pure *Aspergillus niger* culture has been isolated from a green nail. This was the result of an onycholysis caused by a foreign body, in this case, hair in a patient who was a barber (Plates 22 and 23).

2. Candida

Moore and Marcus (1951) claim that *Candida albicans* and

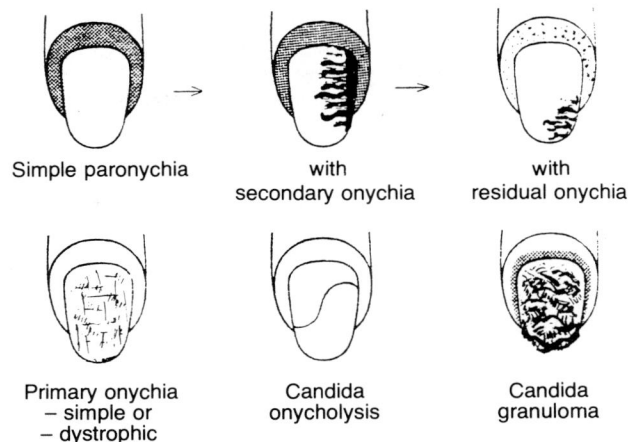

Figure 6.3
Candida onychomycoses.

tropicalis are equally important as causes of the green nail syndrome. *Candida* infections are considered in three groups (Fig. 6.3):

(a) paronychia with secondary onychia (Plate 24) or even subsequent onychia (when the paronychia appears to have healed); the nail plate shows an elongated zone of dark blackish-green discolouration, adjacent to the lateral fold and crossed by transverse grooves
(b) plain *Candida* onycholysis (Plate 25), the frequency of which is underestimated.
(c) primary onychia, simple or dystrophic (rare) (Plate 26); for simplicity candida granuloma is classed with this group.

3. Pseudomonas aeruginosa

Blue–green or black–green nails tend to make one think of Pseudomonas (Plate 27). This nail infection is very similar to the previous ones. The onycholytic type of onychia is particularly interesting because, whereas *Candida* tends to attack the subungual keratin especially in a sugar medium, *Pseudomonas* is capable of keratinolysis, explained by the presence of many trypsin-like (proteolytic) products which are most active in an alkaline medium. In onycholysis, the proteolytic properties of the *Pseudomonas* are demonstrated by their effect on attachment of the nail plate.

Pseudomonas is usually considered an opportunist agent despite rare observations, like those of Shellow and Koplon (1963), where horizontal, parallel green bands marked the nails after paronychial infections. Zuehlke and Taylor (1970) have claimed that *Proteus mirabilis* may be the sole cause of the brownish colour of some nails, and the subsequent presence of *Pseudomonas* plays no part in this discolouration. In our opinion the green nail syndrome is, in practice, usually due to *Pseudomonas* because:

(a) green nails due to *Aspergillus* have not been mentioned in the literature in the last quarter of a century
(b) there is no good biological reason why *Candida* should produce green colouration of nails; in addition species other than *Candida albicans* and *tropicalis* are sometimes found

(c) on the other hand, it is well known that *Pseudomonas* secretes four types of pigment, of which two are important: a blue pigment, pyocyanin, and a green one, pyoverdin. However, some strains yield no pigment. The analysis of metabolic products is often used to distinguish species of bacteria. A pyocyanin-producing bacillus can be tested for the production of pyoverdin, using UV light. This may be useful in diagnosis. It causes a green fluorescence, but this is not diagnostic since the fluorescence can have other causes, such as therapeutic agents, fungi or even bacteria (other species of *Pseudomonas*).

The two main pigments are both water soluble, but only the pyocyanin characteristic of the species, *Pseudomonas aeruginosa*, dissolves in chloroform to give a blue–green colour (Bauer and Cohen, 1957). The practical applications are obvious (Plate 28).

The diagnostic conclusion of the green nail syndrome is summarised in Figure 6.4.

(i) There is a pure strain of *Pseudomonas* so further tests are unnecessary

(ii) The test gives rise to three possibilities:
the culture is sterile
Candida is solely present
culture shows a mixed infection of *Candida* and *Pseudomonas*.

	Cultures	Solubility test
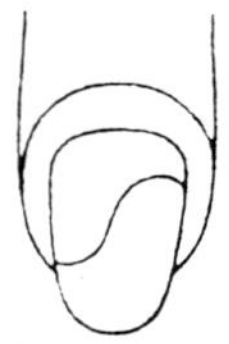	Pseudo-monas	pointless
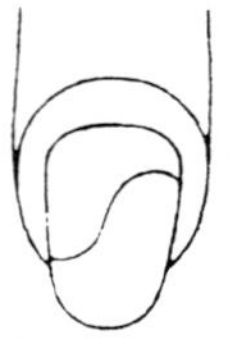	O	yes
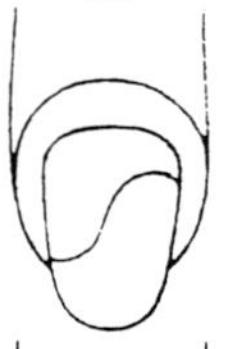	Candida albicans	yes
	Yeast–bacterial association	yes

Figure 6.4
Practical diagnosis of a green-nail syndrome.

If in these three cases, the solubility test is positive,* the presence of *Pseudomonas*, currently or in the past, is indicated for three reasons:

(a) The absence of *Pseudomonas* from cultures does not exclude it from an aetiological role, as shown by Hall *et al*. In *Pseudomonas* infection, the colouration of the nail usually persists for several months despite treatment, whether the culture is positive or sterile.

(b) *Pseudomonas* is surprisingly vulnerable on the skin and nail apparatus. For example, in a case of subungual bulla in green nails, culture of *Pseudomonas* became negative within eight days of the first culture. Probably because hospitalisation of the patient protected it from the effects of moisture.

It is now known that the skin possesses its own stable bacterial ecosystem. Marples has shown that occlusion of the skin leads to an increase in the number of bacteria from less than 4000 to more than 20 000 000 per cm² in 48 hours. Over the second and third days there is an increase in the number of Gram negative bacteria, which then becomes stable. Moisture may be required for colonisation but the normal dry skin should not tolerate infection by *Pseudomonas* or other Gram negative bacteria. In fact, the relative dryness of the skin no doubt explains, in part, the supposed antibacterial properties of the skin.

(c) Using a new selective culture medium, developed by J. F. Vieu at the Pasteur Institute (Paris), we have been able to obtain *Pseudomonas* from an aqueous solution, which had been discoloured by a 'sterile' green nail four years previously.

PARASITIC PARONYCHIA (Plate 29)

The female jiggar flea usually lodges in the skin of the feet and particularly in the interdigital spaces and around the nails. The disease is widespread among the Pygmies, who seem to ignore it. Basset (1973) has observed it in Europeans, who were unaware of the cause of the web and periungual involvement.

BIBLIOGRAPHY

[1] ACHTEN G. — Les onychomycoses. *Bull. Soc. franç. Derm. Syph.*, 1964, *71*, 579–584.
[2] BADILLET G. — A propos de 1258 souches de dermatophytes isolées à Paris de 1956 à 1964. *Presse Med.*, 1966, 74, 973–976.
[3] BADILLET G. — Les dermatophytes de l'enfant. *Ann. Pediat.*, 1969, *44*, 2733–2743.
[4] BARLOW A.-J.-E., CHATTAWAY F.-W., HOLGATE M.-C., ALDERSLEY T. — Chronic paronychia. *Brit. J. Derm.*, 1970, 82, 448–453.
[5] BASSET A. — Dermatoses parasitaires tropicales. *Concours méd.*, 1973, 95, 3648–3661.
[6] BAUER M.-F., COHEN B.-A. — The role of pseudomonas aeruginosa in infection about the nails. *Arch. Derm.*, 1957, 75, 394–396.
[7] GRUPPER C., AVRAM A. — Trichophytie diffuse à trichophyton rubrum chez un malade atteint de psoriasis. *Bull. Soc. franç. Derm. Syph.*, 1972, 79, 610–612.
[8] LEYDEN J.-J., DECHERD J.-W., GOLDSCHMIDT H. — Exfoliative cytology in the diagnosis of psoriasis of nails. *Cutis*, 1972, *10*, 701–704.

*If the solubility test is negative psoriasis has to be borne in mind.

[9] LIAUTAUD B., GROSSHANS E., BASSET M. — Les onychomycoses à champignons saprophytes. *Presse méd.*, 1971, *79*, 1163–1166.

[10] MOORE M., MARCUS M.-D. — Green nails: Role of Candida and Pseudomonas aeruginosa. *Arch. Derm. (Chic.)*, 1951, *64*, 499–505.

[11] SHELLOW W.-V.-R., KOPLON B.-S. — Green striped nails: chromonychia due to Pseudomonas aeruginosa. *Arch. Derm. (Chic.)*, 1963, *97*, 149–153.

[12] STONE O.-J., MULLINS J.-F. — Experimental studies on chronic paronychia. *Arch. Derm. (Chic.)*, 1964, *89*, 455–460.

[13] STONE O.-J., MULLINS J.-F. — Chronic paronychia in children. *Clin. Pediat.*, 1968, 7, 104–107.

[14] ZAIAS N. — Onychomycosis. *Arch. Derm.*, 1972, *105*, 263–274.

[15] ZAIAS N. — Superficial white onychomycosis. *Sabouraudia*, 1966, *5*, 99–103.

[16] ZAIAS N., OERTEL I., ELLIOT D.-F. — Fungi in toe-nails. *J. invest. Derm.*, 1969, *53*, 140–142.

[17] ZUEHLKE R.-L., TAYLOR W.-B. — Black nails with Proteus mirabilis. *Arch. Derm. (Chic.)*, 1970, *102*, 154–155.

7. THE NAIL IN DERMATOLOGICAL DISEASE

R. Baran

PSORIASIS

In established psoriasis the nail changes are identical to those of the skin and the proximal nail fold in particular is often involved. The changes are of two main types: damage to the nail, the nature of which depends on the part of the matrix affected and the duration of the attack; anomalies of the nail bed and hyponychium (Figs. 7.1, 7.2 and 7.3 and Plate 30).

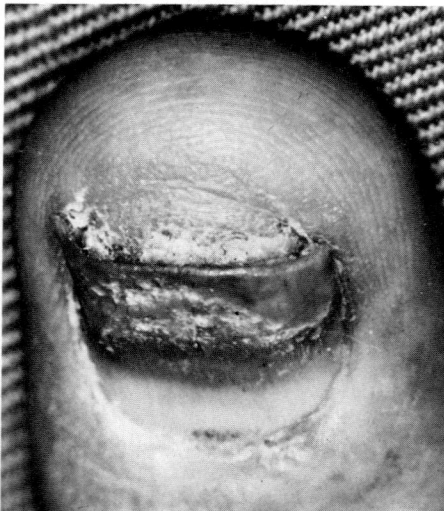

Figure 7.2
Onycholysis with subungual hyperkeratosis (same finger).

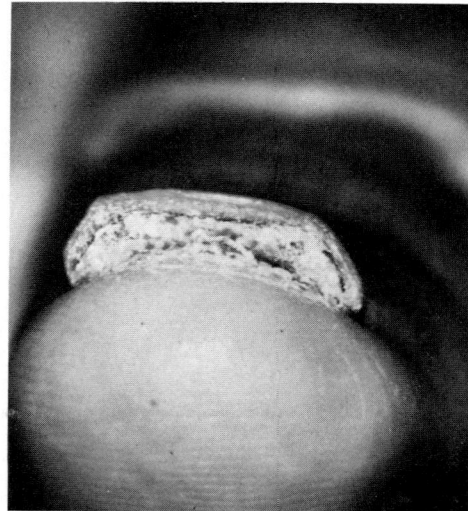

Figure 7.1
Onycholysis with subungual hyperkeratosis.

may be caused by lichen planus (Marks and Samman, 1972; Ronchese, 1965). Clinical suspicion should be confirmed histologically (Zaias, 1967, 1970). Early diagnosis and treatment may then prevent progression towards atrophy and scarring (Table 7.2 and Plate 32).

In chronic psoriasis the nail is only partly affected (apart from certain cases associated with arthropathy). In the acute form the inflammation of the terminal segment is accompanied by changes ranging from deformation to complete disappearance of the nail (Fig. 7.4 and Plate 31). The nail is replaced either by a loosely attached semi-opaque, keratinised crust or hyperkeratotic scales which are firmly attached to the nail bed (Table 7.1) (Lewin et al., 1972; Samman, 1972; Zaias, 1969).

Incidentally De Graciansky et al., have drawn attention to intraungual opaque lines visible on X-ray, probably a result of local treatment.

LICHEN PLANUS

It is essential to be able to recognise the range of lesions which

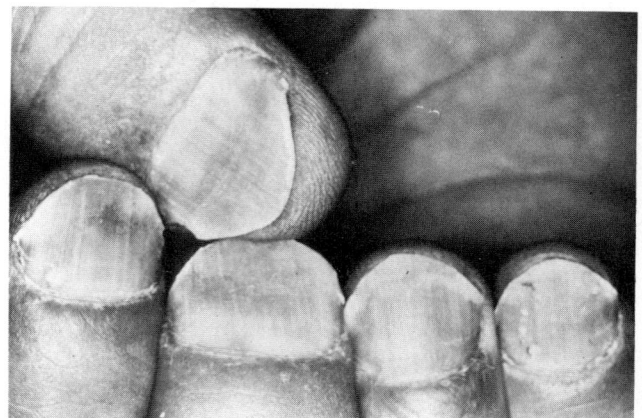

Figure 7.3
Onycholysis with subungual hyperkeratosis tending towards koilonychia.

TABLE 7.1 The nail in psoriasis (modified from N. Zaias, 1969)

Site	Characteristics	Clinical appearance
Matrix	proximal part	isolated pits
	whole area	scattered pock-marks, but also
	proximal and middle parts	transverse furrows and/or leukonychia with a roughened surface
	middle and distal parts	leukonychia with a smooth surface
Nail bed and hyponychium	haemorrhage of dermal papillae of the nail bed	splinter haematomas
	psoriatic changes of the nail bed	reddish, oval spot
	psoriatic changes of the nail bed and hyponychium	onycholysis with a yellow ring surrounding the lesion once the nail starts to loosen, from the central subungual keratosis
	Secondary invasion by *Pseudomonas* and/or yeasts (*Candida*)	yellow-greenish material produced in the area of onycholysis

TABLE 7.2 Lichen planus (modified from N. Zaias, 1970)

Site	Clinical appearance	Frequency (10%)	Prognosis
Matrix	longitudinal furrows alternating with normal nail which appears as ridges	+ +	reversible if attack is not severe
	uniform thinning of the nail	+ +	may be permanent
	pterygium±	+	Permanent
	loss of nail with total nail area atrophy	±	permanent
	longitudinal pigmented band on the nail	±	reversible
Nail bed and hyponychium	purplish lines or papulae may be seen through transparent nail.	+	sometimes reversible
	severe inflammatory reaction with subungual hyperkeratosis, onycholysis or koilonychia	+	sometimes reversible

ULCERATIVE LICHEN PLANUS

This type of lichen planus has been studied by Degos and Schnitzler (1967). The disease causes acropathy of the forefeet and, more rarely, of the hands. The involvement of the nails leads eventually to their disappearance (Cornelius and Shelley, 1967). Transient bullae may be followed by ulcerative lesions which may be haemorrhagic. Often these erosions appear on a surface that is already atrophic and discoloured and the disease rarely involves healthy skin (Figs. 7.5 and 7.6). This chronic and particularly painful disease gives rise to much disability. Lesions of the toe-nails usually coexist with other manifestations of lichen planus.

LICHEN STRIATUS

In some cases the linear papular eruption involves the hand and may extend along the finger as far as the proximal nail fold. A few weeks later a longitudinal depression can be seen in the corresponding part of the nail and eventually ridging, splitting, shredding, onycholysis and total nail loss occurs. The lesion heals after six months (Owens, 1972).

IDIOPATHIC ATROPHY OF THE NAILS

Idiopathic atrophy of the nails (Samman, 1969) is part of a

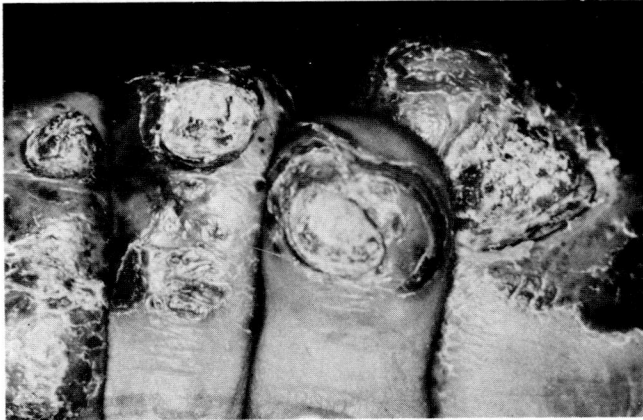

Figure 7.4
Psoriatic arthropathy with the nail replaced by a loosely adherent keratinised crust.

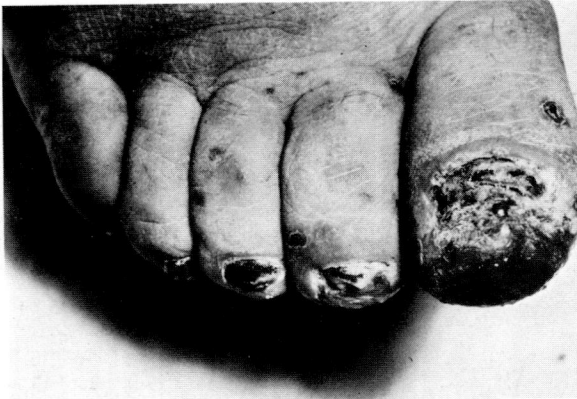

Figure 7.5
Ulcerative lichen planus of the toe-nails; early stage (Degos' collection).

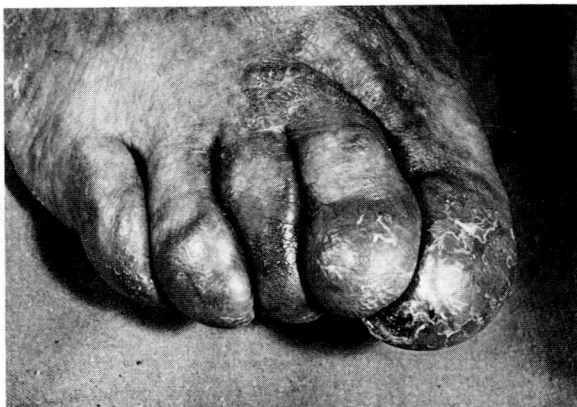

Figure 7.6
Ulcerative lichen planus; late stage.

syndrome rather than a disease (Achten, 1972). The most common associated disorder is akin to lichen planus.

DARIER'S DISEASE

Although the nails are commonly affected by this disease they are rarely involved in isolation (Plate 33 and Fig. 7.7) (Ronchese, 1967; Samman, 1972). The most important characteristics have been detailed by N. Zaias (1973) and are represented in tabular form (Table 7.3). These include nail fragility with thinning, striation, distal fissuring and roughness of the plate which may occasionally be greatly thickened and have a V-shaped notch distally.

Figure 7.7
Darier's disease demonstrating the pathognomonic triad: longitudinally white lines, longitudinal red lines, and distal subungual wedge-shaped keratosis. (Ronchese's collection.)

The pathognomonic triad of clinical features consists of distal subungual keratoses, white and red longitudinal bands and multinucleated giant cells especially present in the nail bed on histological examination (Plate 33 and Fig. 7.10). The nail changes may be the first manifestation of the disease.

ECZEMA OF THE NAIL APPARATUS

There are three types of eczematous involvement of the nail apparatus.
1. Eczema of the fingers with a periungual distribution producing changes in the corresponding nails.
2. Eczema unassociated with periungual involvement, in which the nature of the condition can be established only by recognising the presence of eczematous lesions elsewhere.

In these types, the nail plate becomes thickened, rough with pitting, ridging and transverse furrows, which, when deep enough, are liable to cause a nail to become detached.

Superimposed on these changes, may be those caused by

TABLE 7.3 The nail in Darier's disease (modified from N. Zaias, 1970)

Clinical manifestations	Site or origin
Red longitudinal subungual streak (early stage)	nail bed frequently extending to the hyponychium and lunula
White longitudinal subungual streak (late stage)	nail bed often extending to hyponychium and lunula
Wedge-shaped distal subungual keratosis	distal nail bed and hyponychium
Splinter haemorrhages	region of nail bed subepithelium
True leukonychia	matrix
Flat keratotic papules on the proximal nail fold	proximal nail fold, but also the epidermis covering the pulp

detergents (Fig. 7.11) (Achten, 1963) involving also the cuticle with secondary paronychia.

3. Eczema of the nail bed (Rein and Rogin, 1950). This is usually associated with a wide range of cosmetic products, such as base coats, nail hardeners, false nails, and, more recently, hair setters, which affect the hands of hairdressers. All subsequent lesions, clinically similar, appear over a period of hours, days or even weeks as splinter haemorrhages, soon followed by the development of subungual hyperkeratosis of varying degree. Onycholysis and paronychial infections may finally develop. Colour changes may range from a bluish-red appearance initially to rust, and finally yellow. This type of eczema may occasionally be intensely painful.

Tulip bulbs have recently been recognised as a source of allergen producing onychopathy in horticulturists, who come in constant contact with them. The painful dermatitis particularly involves the thumb, index and middle fingers, starting under the free edge of the nail then extending to the finger-tips and periungual region. The changes are those of a dry eczema with erythema, scaling and fissuring, with a marked degree of subungual hyperkeratosis. The allergen seems to be concentrated in the epidermis of the bulb scales immediately beneath the dead outer covering. It has been identified as lactone and the diagnosis can be confirmed by patch tests with a dilution of 10^{-5}.

Dermatitis caused by other bulbs (hyacinth and narcissi) is more often due to irritation by raphid cells with bundles of needle-shaped crystals of calcium oxalate, and is close to dermatitis caused by fibre-glass.

Escavenitis (Montel and Gouyer, 1957) can be included under a heading of nail-bed eczema. The coelomic fluid of a sea-worm (*Nesreis diversicolor*), used as bait, can cause onychopathy of the first three fingers of the right hand in fishermen. The nails become loosened and a yellow fluid appears under and around the nail. The abnormal mobility of the nail causes pain on the slightest contact.

HALLOPEAU'S ACRODERMATITIS AND ACROPUSTULOSES

The vesicular form of Hallopeau's acrodermatitis (Figs. 7.8 and 7.9) must be distinguished from eczema of the terminal segments. These lesions usually occur in the pulp or in the dorsal aspect of the terminal phalanx. The vesicles are few and discrete and break down to be replaced by crusts. The vesicles tend to reappear within the area first affected.

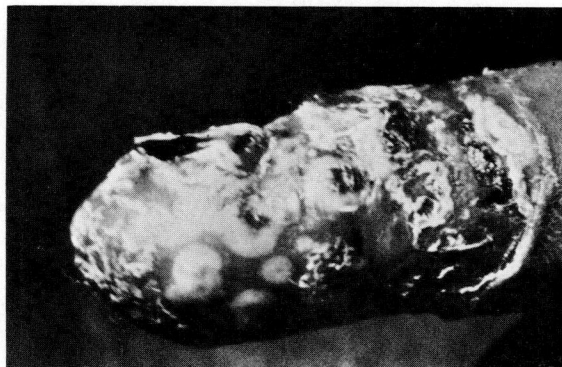

Figure 7.8
Hallopeau's acrodermatitis (Colomb's collection).

The suppurative form develops into a painful paronychial infection, and pus can be extracted. Suppuration can extend to the palmar aspect of the pulp. Areas of suppuration appear either as discrete pustules or as irregular confluent lakes. These pustules break down after a few days leaving exposed, raw dermis, but without true ulceration. They later become covered with crusts and dry scales (Degos, 1976).

Some forms may develop serious functional damage as a result

Figure 7.9
Acropustulosis.

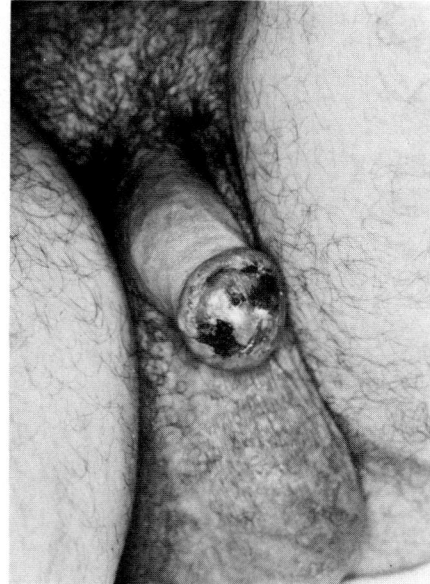

Figure 7.11
Balanitis of same patient (Schnitzler's collection).

of sclerosis and atrophy of the terminal segments with bone resorption. After several attacks, the fingers become tapered, atrophied, the nails are lost and a flexion deformity develops.

No specific organism has been identified and it is necessary to take the general clinical manifestations of the patient into account in order to distinguish it from Hallopeau's pustular acrodermatitis, impetigo herpetiformis, pustular psoriasis, and Reiter's disease (Figs. 7.10 and 7.11).

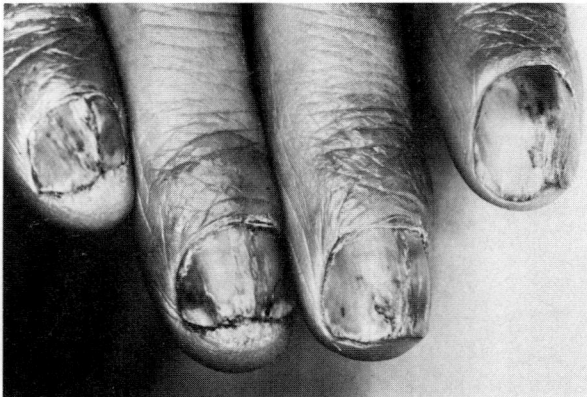

Figure 7.10
Reiter's syndrome (Schnitzler's collection).

DIGITAL HERPES

Herpetic whitlows may affect the terminal phalanx and produce a transient, intensely painful inflammation. In doubtful cases, the so-called 'ballooning degeneration' malpighian cells are diagnostic.

HJORTH–SABOURAUD'S PUSTULAR PARAKERATOSIS

Bacterial parakeratosis of the fingertips has been described by Sabouraud (1931), and Hjorth and Thomsen (1967). It is an eczematous disease involving one or two fingers, usually the thumb and index, rarely more. It usually occurs in girls about the age of seven as papery, erythemato-squamous lesions on the pulp and around the nail. Pustules are only seen in the initial stage. Subungual hyperkeratosis may be associated with the dermatitis and cause deformity of the nail plate.

Although it is often confused with atopic dermatitis, mycosis or psoriasis, Hjorth–Sabouraud's syndrome appears to be a distinct clinical entity.

PEMPHIGUS

On the dorsal aspect of the distal digit, the bulla breaks down leaving an inflamed and ulcerated paronychia. If it occurred in isolation, the diagnosis would be difficult, but its nature can be established histologically and by Tzanck's smear, cytofluorescence and cyto-immunofluorescence (Temime and Cotte, 1973). Following the periungual inflammation of pemphigus, the nails may change in colour and texture with splinter haemorrhages, onycholysis, or even onychomadesis. Shedding of the nail due to nail bed involvement is uncommon (Baumal and Robinson, 1973).

In the case of foliaceous pemphigus the nail is pushed up and eventually falls off, for the same reason as in the majority of exfoliant erythrodermas.

SYPHILIS

This disease is not always easy to recognise. Chancre on the finger, which may be occupational, may form a tumourlike mass, or may be accompanied by painful inflammation of the fingertip, resembling a whitlow. In some cases a crusty ulceration covers the free edge of the nail in a half-moon shape, or may develop in one of the lateral borders of the nail. This primary lesion is accompanied by an epitrochlear or axillary lymphadenopathy.

In secondary syphilis a wide variety of changes may develop, including onycholysis, pachyonychia, pitting, and discolourations (Plate 34). Papulae may be found around the nail and may be keratotic, inflamed (but painless) or ulcerated with superadded infection.

None of these lesions is pathognomonic and diagnosis rests on the other clinical or serological features of the disease.

none the less interesting (Figs. 7.12, 7.13 and 7.14). Some, such as koilonychia, may precede the porphyria (Puissant *et al.*, 1971) others appear as distal hemitorsion. Mutilating porphyria, especially in its toxic forms, destroys the underlying bone and, in consequence, the nail may be reduced to a stump (Perrot, 1968). This also occurs in congenital porphyria. In erythropoietic protoporphyria the nails are opaque, blue–grey or brownish and the lunulae are often absent (Thivolet *et al.*, 1968).

TABLE 7.4 Total Onychodystrophies in late cutaneous porphyria

	Mutilating form	Koilonychia	Distal hemitorsion	'Grotto' nail
H. Perrot (1968)	+			
A. Puissant *et al.* (1971)		+		
A. Popescu *et al.* (1972)		+		
A. Bazex *et al.* (1973)		+		
J. Sayag *et al.* (1975–7)		+		
R. Baran (1977)		+		+
R. Baran (1977)			+	
L. Schnitzler *et al.* (1977)		+		

PORPHYRIA

The nail may show many different changes of colour or morphology in porphyria. The serosanguinous bullae beneath the nail, usually under its distal part, may produce marked discolouration, especially when fungal or *Pseudomonas* infection is present (Plate 35). Other manifestations have already been mentioned, particularly disappearance of the lunula.

Global onychodysmorphias (Table 7.4) are less well-known but

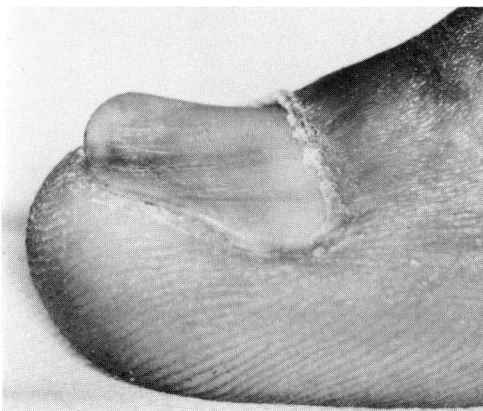

Figure 7.12
Koilonychia.

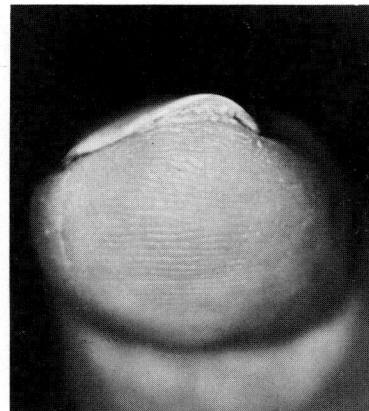

Figure 7.13
Distal hemitorsion.

ALOPECIA AREATA

Authors disagree as to the frequency with which the nails are affected in this disease, ranging from 7 per cent (Muller and Winkelmann) to 66 per cent (Klingmuller and Reeh).

Classically, the severity of the onychopathy is considered to be proportional to the severity of the hairy involvement, especially at its onset. However, severe nail lesions have been observed in benign alopecia and need not imply a poor prognosis. The nail

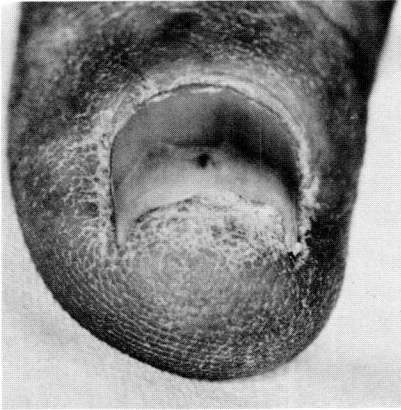

Figure 7.14
Distal arched nail

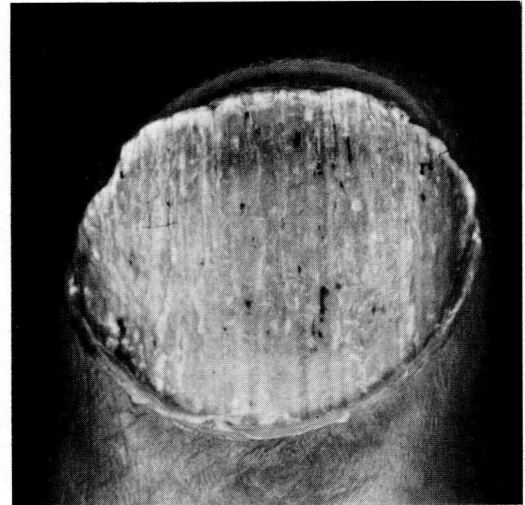

Figure 7.15
Vertical striated sandpaper nail in alopecia areata.

changes involve the form, colour and consistency of the nail.

Koilonychia and severe shortening of the nail have been observed. The nail may also fall off as a result of onychomadesis. In superficial dysmorphia, the longitudinal lines become more marked with onychorrhexis. Development of the horizontal fissures (Plate 36) of Beau's lines may occur. Commonly, pitting appears, which varies from small pits, often less marked than those of psoriasis, to Sabouraud's faceted nail (Plate 37). Sometimes colour changes occur in alopecia, which make it opaque, yellow, brown or grey. Horizontal streaks of leukonychia (Fig. 7.15) can be present. Inflammatory forms rarely occur with deep erythema of the proximal third or of the lunula alone but spotted lunula can be seen.

The nail usually becomes thicker than thinner; its growth is slowed; it is roughened, fragile and friable in the 'vertical striated sandpaper nails' (Baran and Dupré) one of the appearances pertaining to the 'twenty nail dystrophy' (Fig. 7.16).

In the triad, 'Alopecia universalis, onychodystrophy and total vitiligo', the onychodystrophy has the appearance of longstanding fungal infection (Demis and Winer, 1963).

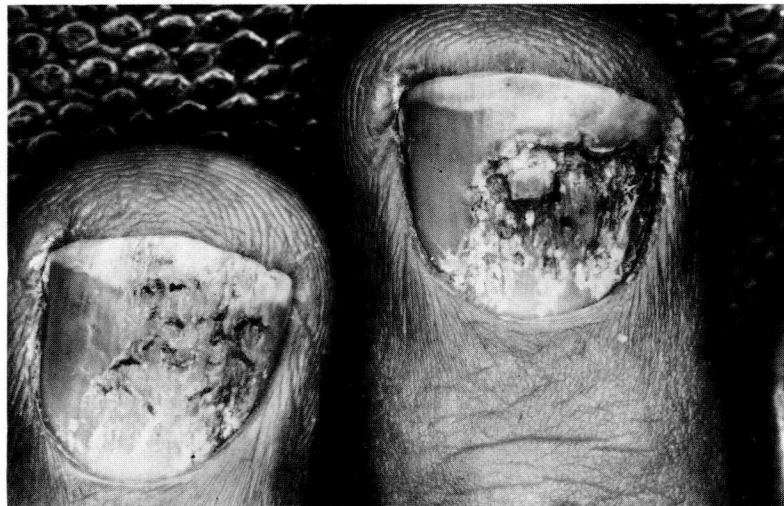

Figure 7.16
Friable nail in alopecia areata.

BIBLIOGRAPHY

PSORIASIS

[1] LEWIN K., DEWIT S., FERRINGTON R.-A. — Pathology of the finger nail in psoriasis. *Brit. J. Derm.*, 1972, *86*, 555–563.
[2] SAMMAN P. — *The nails in disease.* London, W. Heinemann. Med. Books, 1972, 2e édit.
[3] ZAIAS N. — Psoriasis of the nail. *Arch. Derm.* (*Chic.*), 1969, *99*, 567–579.

LICHEN PLANUS

[4] ACHTEN G., WANET-ROUARD J. — Atrophie idiopathique des ongles et lichen plan. *Arch. belg. Derm. Syph.*, 1972, *28*, 251–254.
[5] CORNELIUS C.-E., SHELLEY W.-B. — Permanent anonychia due to lichen planus. *Arch. Derm.* (*Chic.*), 1967, *96*, 434–435.
[6] DEGOS R., SCHNITZLER L. — Lichen érosif des orteils. *Ann. Derm. Syph.* (*Paris*), 1967, *94*, 241–255.
[7] MARKS R., SAMMAN P.-D. — Isolated nail dystrophy due to lichen planus. *Trans. St. John's Hosp. derm. Soc.* (*Lond.*), 1972, *58*, 93–97.
[8] OWENS D.-W. — Lichen striatus with onychodystrophy. *Arch. Derm.* (*Chic.*), 1972, *105*, 457–458.
[9] RONCHESE F. — Nail in lichen planus. *Arch. Derm.* (*Chic.*), 1965, *91*, 347–350.
[10] SAMMAN P.-D. — Idiopathic atrophy of the nails *Brit. J. Derm.*, 1969, *81*, 746.
[11] ZAIAS N. — The nail in lichen planus. *Arch. Derm.* (*Chic.*), 1970, *101*, 264–271.
[12] ZAIAS N. — The longitudinal nail biopsy. *J. invest. Derm.*, 1967, *49*, 406–408.

DARIER

[13] RONCHESE F. — The nail in Darier's disease. *Arch. Derm.* (*Chic.*), 1965, *91*, 617–618.
[14] SAMMAN P.-D. — *The nails in disease*, p. 65–66. London, W. Heinemann, Medical Books, 1978, 3e édit.

[15] ZAIAS N., ACKERMAN A.-B. — The nail in Darier-white disease. *Arch. Derm.* (*Chic.*), 1973, *107*, 193–199.

ECZEMA

[16] ACHTEN G. — L'ongle normal et pathologique. *Dermatologica*, 1963, *126*, 229–245.
[17] DEGOS R. — *Précis de dermatologie.* Paris, Ed. Méd. Flammarion, mise à jour 1976.
[18] REIN G., ROGIN J.-R. — Allergic eczematous reactions of the nail bed due to «under coats». *Arch. Derm.* (*Chic.*), 1950, *61*, 971–983.

PEMPHIGUS

[19] BAUMAL A., ROBINSON N.-J. — Nail bed involvement in pemphigus vulgaris. *Arch. Derm.* (*Chic.*), 1973, *107*, 751.
[20] TEMIME P., COTTE G. — La cytodermatologie. *Panorama méd.*, 1973, *33*, 1–10.

NORWEGIAN SCABIES

[21] SCHIFF B.-L., RONCHESE F. — Norwegian scabies. *Arch. Derm.* (*Chic.*), 1964, *89*, 236–238.

ALOPECIA AREATA

[22] DEMIS D.-J., WINER M.-R. — Alopecia universalis, onychodystrophy and total vitiligo. *Arch. Derm.* (*Chic.*), 1963, *888*, 195–201.

PORPHYRIA

[23] PERROT H. — *La porphyrie cutanée dite tardive.* Lyon, Simep, 1968.
[24] PUISSANT A., DAVID V., LACHIVER D., AITKEN G. — Formes cliniques atypiques de la porphyrie cutanée tardive. *Boll. Istituto. Derm. S. Gallicano*, 1971, 7, 19–30.
[25] THIVOLET J., FREYCON J., PERROT H., GUIBAUD P., BEYVIN A.-J. — Protoporphyrie érythropoïétique. *Bull. Soc. franç. Derm. Syph.*, 1968, *75*, 829–841.

8. NAIL TUMOURS

R. Baran

MALIGNANT TUMOURS IN THE REGION OF THE NAIL

These are rare and comprise squamous cell carcinoma, Bowen's disease, basal cell carcinoma, metastatic tumours, junctional naevus of the matrix and malignant melanoma. To these can be added paraneoplastic acrokeratosis.

SQUAMOUS CELL CARCINOMA

This tumour is rare and often goes unrecognised because of its benign appearance and deceptive clinical development (Nelson and Hamilton, 1970). It occurs in the thumb, index finger and, less frequently, in the big toe. Clinically, it may present as either a keratotic process, which elevates the free margin of the nail, or a discharge is followed by ulceration of the lateral fold, or as an inflamed paronychia. However it presents, its main characteristic is infection, which is always present, and frequently conceals the underlying cancer. The infection may be associated with a pyogenic granuloma which bleeds on contact. The nail may be deformed and ingrowing, which accounts for the throbbing local pain that may be the presenting symptom.

Radiological involvement of the terminal phalanx is inconstant and late in appearance. The tumour develops very slowly and sometimes several years elapse before it is possible to distinguish the carcinoma. It rarely metastasises, even when the local tumour has developed into a fungating mass. Some authors believe that the condition arises from hyperplasia, secondary to a variety of irritations, which include thermal or infective injury. Occupational exposure to X-rays or resulting from therapy can be an aetiological factor. Chronic arsenic poisoning may act in the same way.

Biopsy is essential for diagnosis, but it may be difficult to differentiate the histology from keratoacanthoma, mainly in *epithelioma cuniculatum*, the symptomatology of which being sometimes very similar. Malignant melanoma in its amelanotic form, or haemangio-endothelio-sarcoma may also enter into the differential diagnosis (Duperrat *et al.*, 1969).

BOWEN'S DISEASE

This is a rare condition; Cosney, Mehregan and Fosnaugh, (1972) have recently reported four cases, as has Mikhail, (1974) while Larregue *et al.*, (1974) have found five others. We have also published five cases recently (Plates 38 and 39) (Baran *et al.*, 1979).

Men are most commonly affected and usually men over 50 years old. The condition frequently occurs in the thumb and index finger of the left hand. The lesion presents as a fissure, which may be painful, in the depths of the lateral or proximal fold with sometimes a whitish cuticle. It gradually extends towards the nail bed and causes separation of the nail by accumulation of abnormal keratin. The progression takes place over three to five years.

At this stage, an erythemato-squamous lesion, fissure or ulcer, or an irregular thickening of the nail bed is discovered. The key of the diagnosis is the histological picture. However, it is not always a simple matter to separate invasive from *in situ* carcinomas beneath the nail.

Differential diagnosis includes periungual warts, onychomycosis, paronychia, eczema, pyogenic granuloma, pyodermatitis, tuberculosis, verruca, subungual exostosis, glomus tumour, keratoacanthoma, amelanotic malignant melanoma and squamous cell carcinoma.

BASAL CELL CARCINOMA

This is extremely rare in the nail. It is characterised by a long period of chronic infection with exuberant subungual granulation. Diagnosis rests on the histology.

METASTATIC TUMOURS

Metastases from visceral cancers in the nail area are extremely rare and usually bronchogenic (Colson and Willcox, 1948; Hammer and Gollmann, 1972). They present as an acute paronychia, which may be pulsatile. However, the rapid development of these lesions contrasts strangely with the continuing good function of the finger. Radiological examination will show involvement of bone. If the primary tumour is unrecognised, biopsy is necessary and will differentiate the lesion from a primary malignant tumour of the bone.

JUNCTIONAL NAEVUS OF THE MATRIX AND SUBUNGUAL MALIGNANT MELANOMA

A junctional naevus of the matrix can, even if benign, produce longitudinal melanotic bands in the nail, but the slightest alteration in these demands immediate biopsy (Plate 40).

It has been believed that the nail matrix does not contain active melanocytes. However, Higashi and Saito (1969) have demonstrated the existence of dopa-positive melanocytes in the matrices of healthy nails of Japanese subjects. In patients with longitudinal, pigmented streaks in the nail, Higashi (1968) did not find a melanoma of the matrix, but simply an increase in the number of melanocytes, with an increase in their dopa-positive activity. The same is true for Negroes and Indians, who

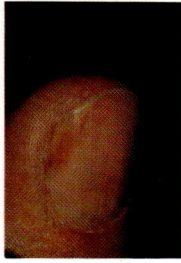

30

31

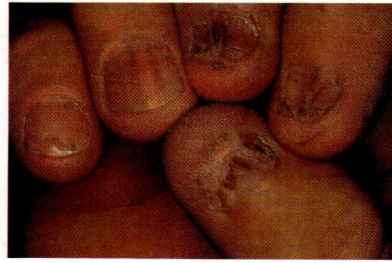

32

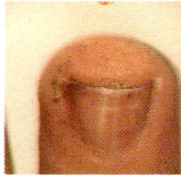

33

34

35

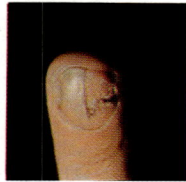

36

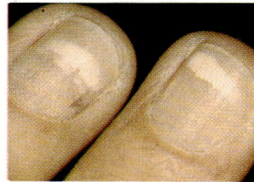

37

Plate 30 Psoriasis at the beginning of an attack. Note the oil stain under the nail

Plate 31 Psoriasis universalis with ungual residue

Plate 32 Lichen planus. All the anomalies are shown here

Plate 33 Darier's disease with its pathognomonic triad : longitudinal white lines, longitudinal red lines, distal subungual keratosis

Plate 34 Syphilitic chromonychia. (Labouche's collection)

Plate 35 Porphyria cutanea tarda. Digital bullae and subungual bullae with pyocyanic graft

Plate 36 Alopecia areata. (Achten's collection)

Plate 37 Alopecia areata. Horizontal striatal leuconychia with trachonychia

38

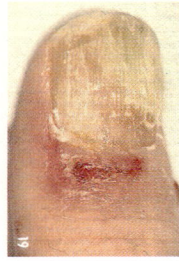

39

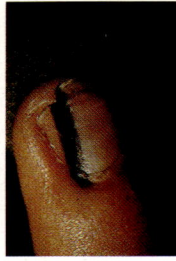

40

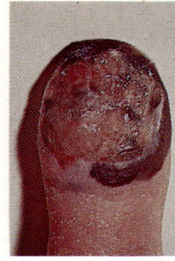

41

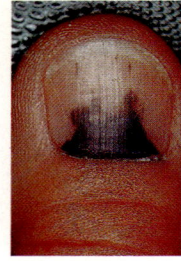

42

43

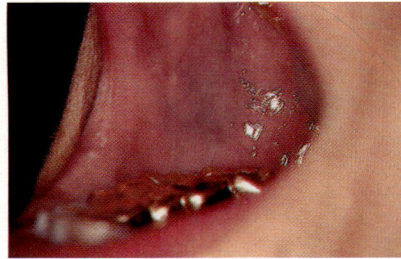

44

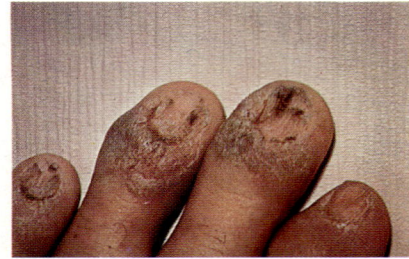

45

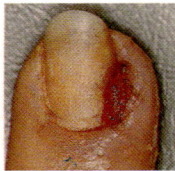

46

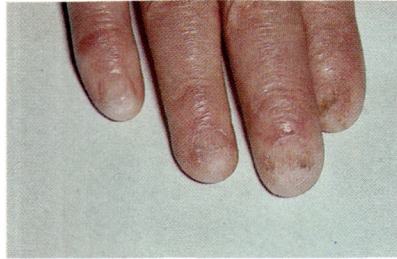

47

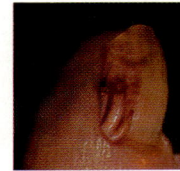

48

Plate 38 Bowen's disease

Plate 39 Bowen's disease

Plate 40 Melanic line (Grupper's collection)

Plate 41 Malignant melanoma of the nail apparatus

Plate 42 Haematoma

Plate 43 Laugier's melanosis

Plate 44 Laugier's melanosis. Same patient

Plate 45 Bazex's and Dupré's paraneoplastic acrokeratosis

Plate 46 Botryomycoma of the type with a pulpy granule on the nail crease

Plate 47 Primary systemic amyloidosis (Civatte's collection)

Plate 48 Koenen's tumour

frequently have melanotic bands or streaks in their nails (Leyden *et al.*, 1972). Similar appearances may occur in white people, in which case they are associated with Laugier's syndrome (Plates 43 and 44).

In France Duperrat *et al.*, (1968) have made an extensive study of subungual melanoma. This starts as a blackish subungual stain usually in the lunula. The development of the lesion distinguishes it from a simple haematoma. The stain of the malignant melanoma grows and the nail splits longitudinally. A brownish vegetating tumour appears which stains dressings brown (Plate 41) and the diagnosis is then obvious. A haematoma, whether it is traumatic or spontaneous, will disappear within a few weeks as the nail grows (Plate 42). A haematoma also appears suddenly, whereas the melanoma develops gradually. Malignant melanoma is most common in the thumbs and big toes. Dufourmentel *et al.*, stress the importance of recognising the spread of pigmentation around the nail in the early diagnosis, which had already been recognised by Hutchinson in his original description.

Diagnosis of malignant melanoma may be further complicated by the effects of trauma:

a haematoma following injury, for example, may conceal the malignant growth

amelanotic forms of tumour may present as a granulation appearing in the centre of the nail

the nail may be progressively destroyed by an exuberant granulation with inconspicuous pigmentation.

It is very rare for lentigo maligna to involve the nail (Lupulescu *et al.*, 1973).

ACROKERATOSIS PARANEOPLASIA OF BAZEX AND DUPRÉ (Bureau *et al.*, 1971)

This condition, which occurs only in men, must be regarded as a paraneoplastic lesion, since its appearance is governed by the existence of cancer, usually of the respiratory or alimentary tract. The condition may disappear if the cancer is removed, but reappears if the cancer returns.

The lesions occur in the extremities and are symmetrical, affecting the hands, feet and rarely the nose and the ears. Involvement of the nail is constant and is the first manifestation. The nails are deformed by subungual hyperkeratosis, which elevates the distal edge. The surface becomes laminated, irregular and whitish in colour. The changes are gross and may resemble major psoriatic onychopathy; in extreme cases the nail will fall off (Plate 45; Fig. 8.1).

In some varieties, atrophy of the nail is total or almost so, the nail bed being replaced by a soft, smooth epidermis, with adherent remnants of the nail plate. If the attack is slight, the nail becomes soft and friable. Sometimes the two types of nail pathology respond to the activity of the underlying tumour and hence there are cases in which the proximal third of the nail is atrophic, while in contrast the distal part is hypertrophic.

The skin lesions covering the terminal phalanges spread on to the backs of the fingers, often accompanied by paronychia with intermittent suppuration. Irregular cracked keratin formations are found at the ends of the fingers and toes.

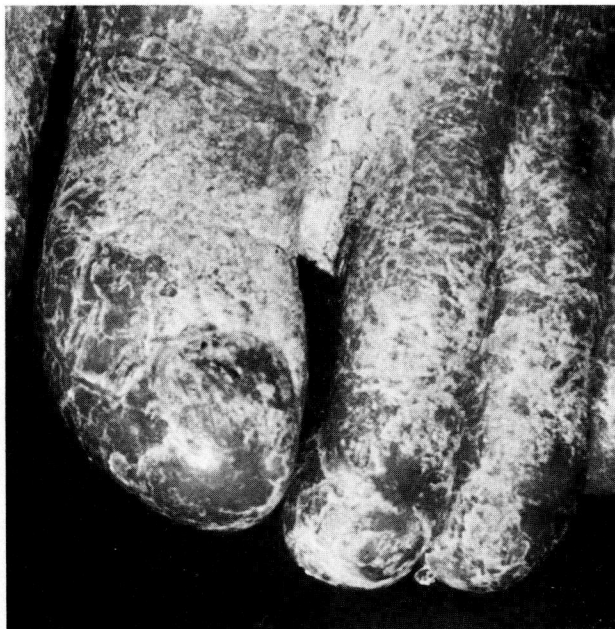

Figure 8.1
Acrokeratosis paraneoplasia (Barrière's collection).

The histological features described by Bazex are inconstant. There is an allergic vasculitis within the dermis, and the epidermal cells show a cytoplasm rapidly becoming eosinophilic and then disappearing and leading to vacuolated cells.

BENIGN NAIL TUMOURS

These are common and may be isolated lesions or part of a general disease. Among the isolated lesions are warts; pyogenic granuloma and ingrowing nail; mucous pseudocyst; glomus tumour; chondroma and subungual exostosis; keratoacanthoma (molluscum sebaceum); subungual epidermoid inclusions; and fibromas.

The general diseases, which will be discussed, are epiloia; incontinentia pigmenti; and primary systemic amyloidosis.

PERI- AND SUBUNGUAL WARTS

These are benign, contagious and auto-inoculable lesions due to a virus. They appear in the periungual grooves as keratotic growths with an irregular surface, frequently cracked. They are not painful but, as they occur commonly in nail-biters, they may be complicated by haemorrhage from irritation, or even infection (Fig. 8.2). They must be distinguished from onychophosis, which is simply a callosity of the nail folds of the toes.

The subungual wart is always painful; it grows under the free edge of the nail and on transillumination appears as a yellow translucent stain (Fig. 8.3). It can be confused with the subungual glomus tumour or very occasionally with tuberculosis verruca.

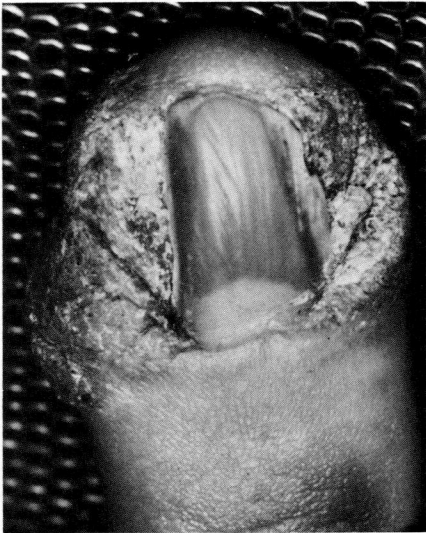

Figure 8.2
Periungual wart with onychodysmorphia.

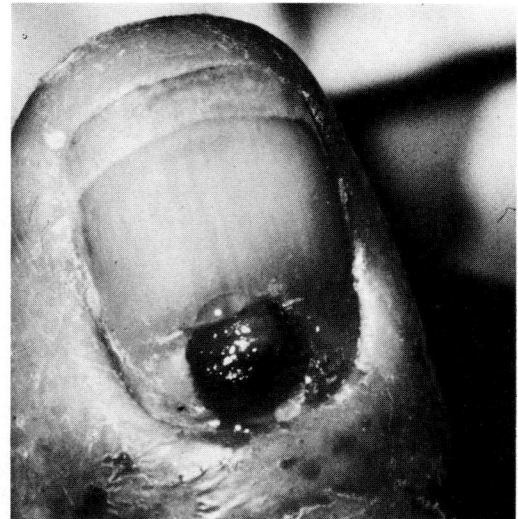

Figure 8.4
Pyogenic granuloma on post-traumatic nail fissure.

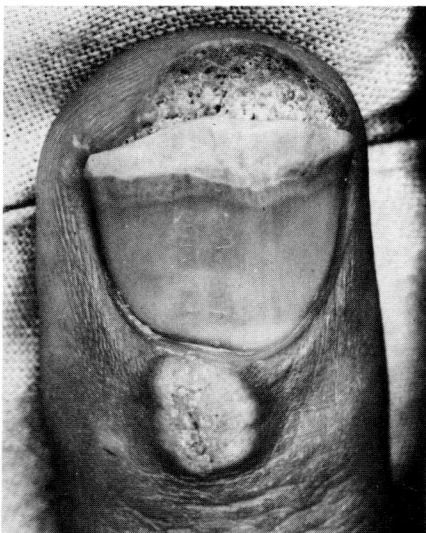

Figure 8.3
Wart under nail plate on and proximal nail fold.

PYOGENIC GRANULOMA

This is a small, vascular, inflammatory, painless tumour of infective origin, the result of an excoriation or sting which may not have been noticed. It is about the size of a pea or larger, dark red in colour, and bleeds readily if touched. It has a rather narrow stalk and a furrow separating it from the healthy skin. It is usually found on the nail folds (Plate 46), but also on a nail that has been cracked by injury (Fig. 8.4). On the big toe, it is usually the result of an ingrowing toe-nail, on the margin of which it appears.

MUCOUS PSEUDOCYSTS

These usually occur in women, more often in the fingers than in the toes. They appear as swellings, 5–12 mm in diameter, round, smooth, with thin overlying skin. Their nature is revealed by transillumination. They are found on the backs of the fingers between the distal interphalangeal joint and the subungual fold (Fig. 8.5), frequently nearer the nail than the joint, and sometimes adjoining the nail plate, in which case a longitudinal groove appears on the plate that disappears when the cyst is removed. About 15 per cent of patients have osteoarthritis and Heberden's nodules of the terminal joint (Arner *et al.*, 1956).

Rarely paronychial fistula may develop on the back of, or underneath, the proximal nail fold (Fig. 8.6). The cysts contain clear viscous fluid. Surgery may often fail without injection of methylene blue in the distal interphalangeal joint because

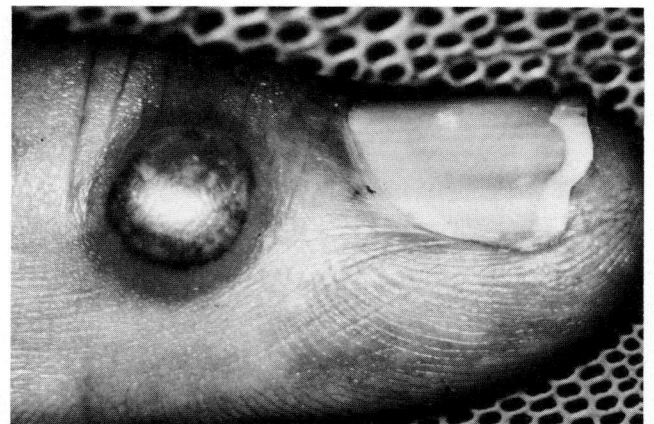

Figure 8.5
Mucoid cyst with longitudinal groove.

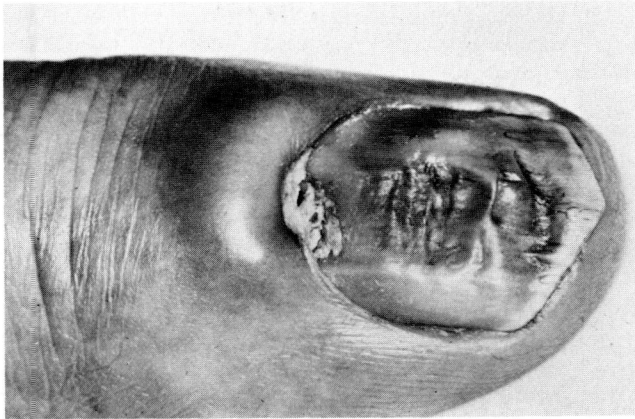

Figure 8.6
Mucoid cyst, paronychial variety.

complete excision is often difficult, due to a lack of clear differentiation between healthy and pathological tissue, and because of communication which may exist with the terminal interphalangeal joint (Johnson *et al.*, 1965). Injection of corticosteroids or hyaluronidase provides only temporary or inconsistant results reducing the activity of the fibroblastic secretion (Dublin, 1972).

GLOMUS TUMOUR

This lesion grows in the nail bed and may be the result of injury. The main characteristic is spontaneous pain, which varies greatly in severity and is usually triggered by pressure: the slightest knock, even just brushing against an object, can induce intense pain, extending up to the shoulder, and resulting, in extreme cases, in disuse atrophy of the hand or even the entire arm. Alteration in temperature may also induce pain. When transilluminated, the tumour appears as a small, bluish stain, a few millimetres in diameter. Sometimes the temperature around the lesion is increased. The nail may be pushed upwards and deformed.

Erosion of the phalanx may be seen on X-ray, due to pressure by the tumour. Occasionally, it may appear as an intra-osseous cyst, in which case the X-ray appearances would not be diagnostic on their own.

Sometimes clinical examination and X-ray do not reveal the tumour and in these cases arteriography may be of value – the tumour showing a star-shaped telangliectatic zone in the arterial phase and confluent lakes in the venous phase (Camirand and Giroux, 1970; Natali *et al.*, 1966).

ENCHONDROMA AND EXOSTOSIS

An enchondroma is a painful benign tumour of the cartilage. If it involves the distal phalanx it may present as a club finger but, if it presents in the region of the proximal nail fold, it may resemble paronychia (Shelley and Ralston, 1964) and it may, like a mucous cyst, produce a longitudinal groove in the nail itself.

Other cases may show a variety of localised nail changes (Yaffee, 1965).

Pathological fractures may occur as a result of bone destruction. Radiologically, the chondroma appears as a cystic defect expanding the bone. Surgical curettage and bone grafting are required.

Subungual exostosis occurs mainly in children, particularly girls, and usually involves the big toe (Fig. 8.7). Aetiology is probably traumatic, but this is disputed. Recently some cases, involving mainly the index finger have been reported (Baran and Sayag).

The lesion is painful, particularly when walking; it elevates the free edge, or lateral margin of the nail, and may cause it to fall off, in which case the underlying tumour may become eroded and infected. An X-ray reveals its nature, showing a dense, sessile or pedunculated osteoma, arising from the terminal phalanx (Cohen *et al.*, 1973). Treatment consists of surgical removal.

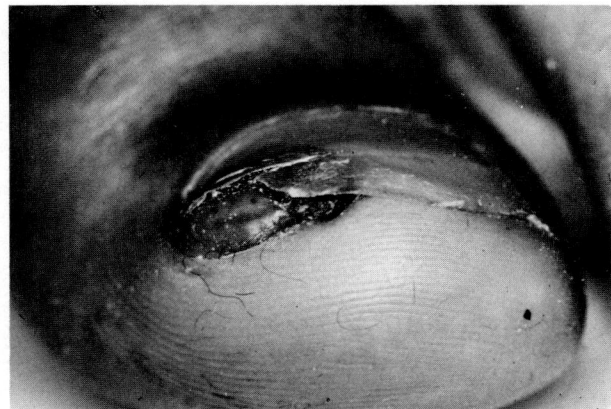

Figure 8.7
Exostosis.

KERATOACANTHOMA (MOLLUSCUM SEBACEUM)

This benign lesion may be confused with squamous carcinoma and has led to unnecessary amputations (Shapiro and Barat, 1970; Mehregan and Fabian, 1973). The eruptive or multiple forms are easy to diagnose, but the isolated lesion may present difficulties. The disease starts as a small inflammatory papule of the thumb or index finger and grows quickly from a height of a few millimetres to 2 centimetres in four to eight weeks. The clinical features consist of pain, swelling and erythema of the distal segment of the finger and of the tissue around the nail (Fig. 8.8). An encrusted nodule usually appears under the free edge of the nail, which may be elevated and eroded (Fig. 8.9).

The lesion may remain static for several weeks and then a very slow regression occurs, leaving a scar. An X-ray will show very early that bone resorption has already occurred, due to compression of the phalanx. The speed of development of the lesion is the key to diagnosis. Surgery should be conservative, but recurrence is not uncommon.

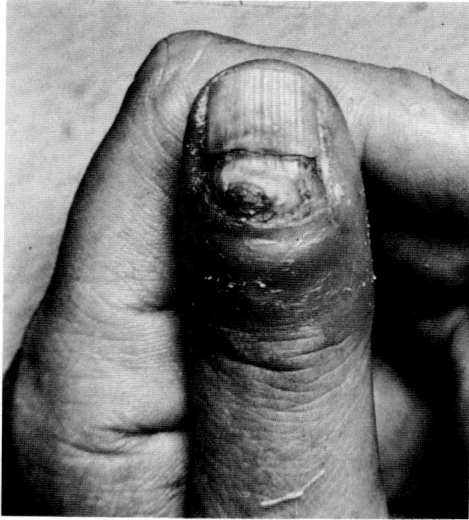

Figure 8.8
Keratoacanthoma of the matrix (Ronchese's collection).

SUBUNGUAL EPIDERMOID INCLUSIONS

Lewin (1969) has described under this title the development of bulbous proliferations of the tips of the rete ridges below the nail plate. Later these may migrate deeply into the dermis of the nail bed. The lesions are usually microscopic, but can be big enough to cause subungual swellings.

The aetiology of this condition is obscure. Subungual epidermoid inclusions are often found in club fingers, but it is probable that they are the result of several different kinds of stimuli including trauma.

NAIL FIBROMAS

These may occur in isolation or can be part of epiloia. They are

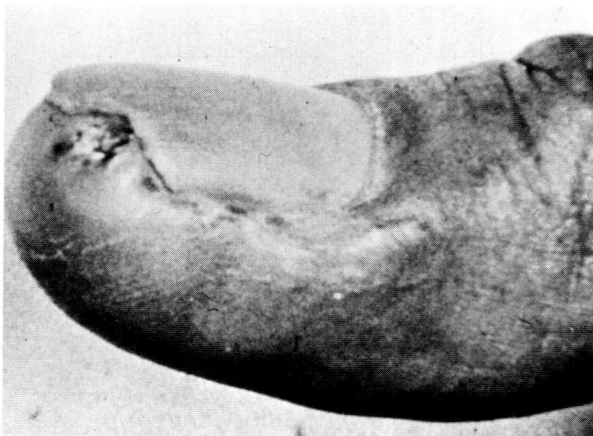

Figure 8.9
Keratoacanthoma under the distal end of the nail (Mehregan's collection).

small, smooth, fibrous tumours which may be lobulated or papillomatous. Frequently they appear under the proximal nail fold and occupy part of a longitudinal groove in the nail. When the extremities become keratotic, they are referred to as Bart's acquired periungual fibrokeratoma. The core of the tumour duplicates the normal dermis. Some forms may be associated with atrophy of the nail.

Steel's garlic-clove fibromas are acquired digital fibro-keratomas. Reye's benign juvenile fibromatosis of the fingers (Fig. 8.10) is a disorder in which hard, hemispherical, smooth

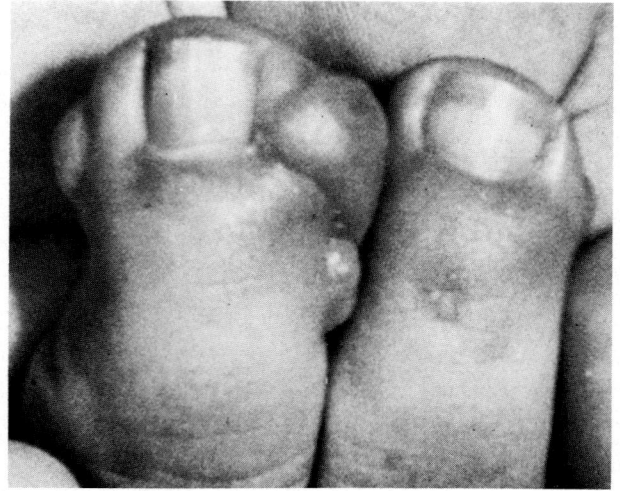

Figure 8.10
Reye's juvenile fibromatosis (Mascaro's collection).

nodules develop, which may be pink or ordinary skin colour. The condition is usually found on the backs or sides of the fingers and the nail may be pushed up, but not destroyed. Histologically there is a proliferation of connective tissue with loss of elastic fibres beneath the thinned epidermis. In electron microscopy, 2 per cent of the fibroblasts show cytoplasmic inclusions, which suggest a dysplasia of elastin.

BENIGN NAIL TUMOURS IN GENERAL DISEASE

Epiloia (tuberous sclerosis). Fibromas have been seen in half of the cases of tuberous sclerosis (Fig. 8.11) and are as frequent as renal hamartomas (Nickel and Reed, 1962). They do not usually appear until puberty and develop as the patient grows older.

Koenen's tumours are one of the most important indications of the disease on the skin (Plate 48). These are delicate grain-shaped proliferations that start in the periungual groove and extend along the back of the nail. Their free end is pointed and slightly keratotic. They may be multiple and present in the form of an arch round the base of the nail.

Sometimes the fibromatous lesions are so big that they cause hypertrophy of the whole nail bed. These exuberant forms are painful when they rub against shoes and must be excised. Recurrence is not unusual. Histologically there are often neural

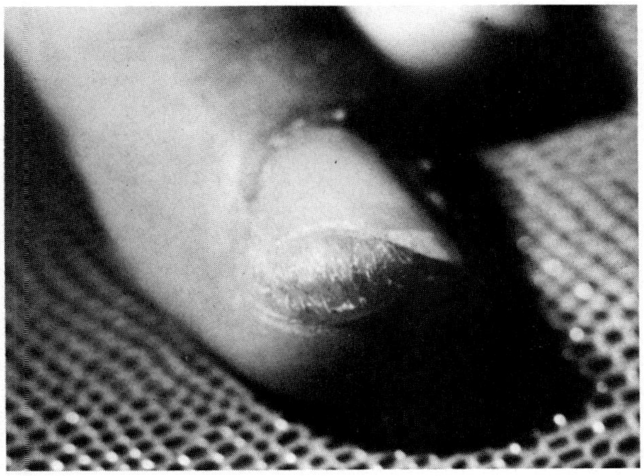

Figure 8.11
Epilbia, subungual fibroma.

elements and giant cells in a connective tissue can sometimes be found. However, according to Verallo the histological picture of Koenen's tumours mimics that of acquired digital fibroma.

Cowden's syndrome is rare but deserves to be mentioned because of its isolated subungual fibrous nodules accompanied by multiple hamartomas.

Incontinentia pigmentia. Hartman (1966) and Pinol *et al.*, (1973) have observed the existence of painful hyperkeratotic tumours under the nail, which appear in adolescents aged between 15 and 20 years and heal spontaneously although they may recur. Clinically and histologically these tumours are lesions that are seen in the verrucous stage of incontinentia pigmenti. In Hartman's case, X-rays demonstrated the erosion of the distal phalanges.

Systemic amyloidosis. Subungual papillomatosis is one of the characteristics of this disease. Profuse corneal vegetations, light brown in colour and of elastic consistency, push up and split the nail. A biopsy will show the presence of amyloid (Duperrat *et al.*, 1955).

Even more common nail changes in this disease are like

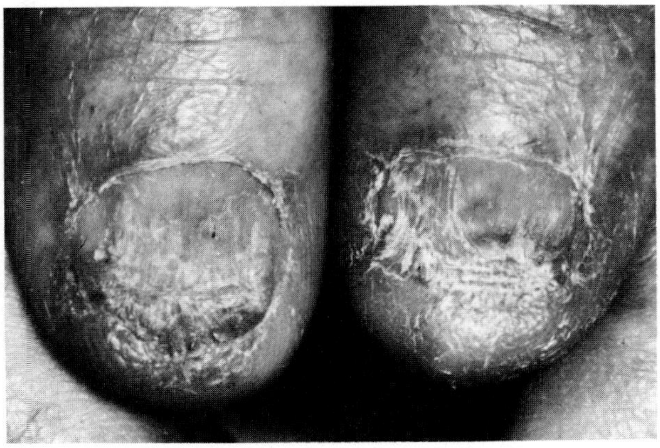

Figure 8.12
Systemic amyloidosis (Civatte's collection).

dystrophic lichen planus with striations, crumbline of the free margin, punctated erosions, fragility and anonychia.

We have seen a case of atrophy of all the finger nails, similar to lichen planus, with a typical amyloidosis histology (case of J. Civatte *et al.*, Plate 47; Fig. 8.12). With J. Sayag, we have seen a bullous primary amyloidosis, which looked deceptively like a porphyria cutaneous tarda, with a slight koilonychia and a longitudinal melanotic band.

BIBLIOGRAPHY

MALIGNANT TUMOURS

[1] NELSON L.-M., HAMILTON C. — Primary carcinoma of the nail bed. *Arch. Derm. (Chicago)*, 1970, *101*, 63–67.
[2] DUPERRAT B., GREPINET H., CHAMBON R. — L'épithélioma du lit unguéal. *Bull. Soc. franç. Derm. Syph.*, 1969, 76, 560–562.
[3] COSNEY R.-J., MEHREGAN A., FOSNAUGH R. — Bowen's disease of the nail bed. *Arch. Derm. (Chicago)*, 1972, *106*, 79–80.
[4] MIKHAIL G.-R. — Bowen's disease and squamous cell carcinoma. *Arch. Derm. (Chicago)*, 1974, *110*, 267–270.
[5] LARREGUE M., CIVATTE J., BELAICH S., LETESSIER S., DEGOS R. — Maladie de Bowen. *Bull. Soc. franç. Derm. Syph.*, 1974, *81*, 24–26.
[6] BARAN R., *et al.*, — Maladie de Bowen de l'appareil unguéal. *Ann. Derm. Vénéréol.*, 1979, *106*, 227–233.
[7] COLSON G.-M., WILLCOX A. — Phalangeal metastases in bronchogenic carcinoma. *Lancet*, 1948, *1*, 101–102.
[8] HAMMER B., GOLLMANN G. — Métastases aux phalanges dans le cancer bronchique. Contribution à la fixation du temps de survie. *Münch. med. Wschr.*, 1972, *114*, 61–63.
[9] HIGASHI N., SAITO T. — Horizontal distribution of dopa-positive melanocyte in nail matrix. *J. invest. Derm.*, 1969, *53*, 163–165.
[10] HIGASHI N. — Melanocytes of nail matrix and nail pigmentation. *Arch. Derm. (Chicago)*, 1968, *95*, 570–574.
[11] LEYDEN J., SPOTT D.-A., GOLDSCHMIDT H. — Diffuse and banded melanin pigmentation in nails. *Arch. Derm. (Chicago)*, 1972, *105*, 548–550.
[12] DUPERRAT B., PUISSANT A., GLICENSTEIN J., CHERIF-CHEIK J.-L. — Le panaris mélanique post-traumatique. *Bull. Soc. franç. Derm. Syph.*, 1968, *75*, 582–584.
[13] LUPULESCU A., PINKUS H., BIRMINGHAM D., USNDEK H., POSCH J. — Lentigo maligna of the fingertip. *Arch. Derm. (Chicago)*, 1973, *107*, 717–722.
[14] BUREAU Y., BARRIERE H., LITOUX P., BUREAU B. — Acrokératose paranéoplasique de Bazex. Importance des lésions unguéales. *Bull. Soc. franç. Derm. Syph.*, 1971, *78*, 79–81.

BENIGN TUMOURS

[1] ARNER O., LINDHOLM A., ROMANUS R. — Mucous cysts of the fingers. *Acta chir. scand.*, 1956, *111*, 314.
[2] JOHNSON X.-C., GRAHAM J.-H., HELWIG E.-B. — Cutaneous myxoïd cyst. *JAMA*, 1965, *191*, 109.
[3] DUBIN H.-V. — Mucinous cysts and mucinous pseudocysts. *Symposium sur les tumeurs cutanées.* Miami, American Academy of Dermatology, 1972.
[4] CAMIRAND P., GIROUX J.-M. — Subungual glomus tumor. *Arch. Derm. (Chicago)*, 1970, *102*, 677–679.
[5] NATALI J., ESCARLAT B., VINARDI G., BATISSE F. — Artériographie d'une tumeur glomique. *J. Chir.*, 1966, *92*, 481–484.
[6] SHELLEY W.-B., RALSTON E.-L. — Paronychia due to enchondroma. *Arch. Derm. (Chicago)*, 1964, *90*, 412–413.
[7] YAFFEE H.-S. — Peculiar nail dystrophy caused by an enchondroma. *Arch. Derm. (Chicago)*, 1965, *91*, 361.

[8] BARAN R., SAYAG J. — Une localisation exceptionnelle: l'exostose de l'index. *Ann. Derm. Vénéréol.* (à paraître).

[9] COHEN H.-J., FRANCK S.-B., MINKIN W., GIBBS R.-C. — Subungual exostosis. *Arch. Derm. (Chicago)*, 1973, *107*, 431–432.

[10] SHAPIRO L., BARAF C.-S. — Subungual epidermoid carcinoma and kerato-acanthoma. *Cancer*, 1970, *25*, 141–152.

[11] MEHREGAN A.-H., FABIAN L. — Kerato-acanthoma of nail bed. A report of two cases. *Intern. J. Derm.*, 1973, *12*, 149–151.

[12] LEWIN K. — Subungual epidermoïd inclusions. *Brit. J. Derm.*, 1969, *81*, 671–675.

[13] STEEL H.-H. — Garlic clove fibroma. *JAMA*, 1954, *191*, 1 082–1 083.

[14] NICKEL W., REED B. — Tuberous sclerosis. *Arch. Derm. (Chicago)*, 1962, *85*, 209–226.

[15] HARTMAN D.-L. — Incontinentia pigmenti associated with subungual tumour. *Arch. Derm. (Chicago)*, 1966, *94*, 632–635.

[16] PINOL AGUADE J., MASCARO J.-M., HERRERO C., CASTEL T. — Tumeurs sous-unguéales dyskératosiques douloureuses et spontanément résolutives. Les rapports avec l'Incontinentia Pigmenti. *Ann. Derm. Syph. (Paris)*, 1973, *100*, 159–168.

[17] DUPERRAT B., CHASSIGNEUX J., SOLCEER R. — Végétations cornées sous-unguéales révélatrices d'un myélome plasmocytaire multiple. *Bull. Soc. franç. Derm. Syph.*, 1955, *62*, 474.

9. NAIL CHANGES IN GENERAL PATHOLOGY

R. Baran

CARDIOPULMONARY DISORDERS

In congenital heart disease (Plate 49), where cyanosis is a feature, the disease appears before the deforming changes in the fingers which are proportional to the degree of cyanosis, reflecting the importance of polycythaemia. For unknown reasons the changes begin in the index and the thumb and lead nails to their increased growth.

The finding of cyanosis and finger clubbing in chronic pulmonary insufficiency indicates anoxia and demands examination for signs of chronic cor pulmonale. In chronic bronchial infections, the appearance of clubbing depends on three criteria: long duration, often more than ten years, hypoxia, and hyperglobulinaemia. If one is missing, there is little likelihood of its developing. In bronchogenic carcinoma, the mechanism is different, as there are no circulatory disturbances and the history is short. Possibly induction of hormones by the tumour may stimulate the opening of arteriovenous shunts in the fingertips.

Terry (1954) has described 'red half moons', where the red colour of the lunulae is of a variable intensity. This differs from the cyanotic nail bed seen in advanced congestive cardiac failure. Aortic insufficiency is sometimes accompanied by a visible capillary pulse in the nail bed or in the skin of the proximal nail fold, although this is not pathognomonic of the condition.

PERIPHERAL VASCULAR DISEASE

To understand the changes that this produces in the nail the following should be appreciated:

(a) ischaemic symptoms can result from prolonged arterial spasm in the absence of any apparent arterial obstruction; nevertheless, as they are often localised in a few of the digits, obstruction in these fingers seems probable (Samman and Strickland, 1962).

(b) clinical manifestations, such as those of Raynaud's disease, may evolve over many decades without causing problems, when the spasm affects only the smallest vessels.

(c) major ischaemic necrosis in the finger is almost always associated with arterial obstruction, but there is no clear correlation between the clinical appearance and the results of paraclinical investigations (Baran et al., 1972).

(d) the characteristics of the lesions are most obvious when several digits are affected and there is a history of longstanding arterial spasm.

(e) abnormality of the ulnar artery combined with that of the superficial palmar arch seems to be of great importance in ischaemic disease.

The nail becomes thin and brittle with longitudinal striae and tends to fissure at the free border. Koilonychia can be seen. The thinning of the nail renders the nail bed more visible and redder than normal, contrasting with areas whitened by onycholysis or darkened by accumulation of debris and secondary infection.

In contrast, mainly in older patients, the toe nail sometimes thickens, obscuring the underlying bed and may, in some instances, develop into onychogryphosis. The appearance of pachyonychogryphosis is characteristic of trophic varicose lesions.

Cold hands often develop chronic paronychia. Transverse depressions of the nail result from nutritional deficiency with temporary arrest of growth. If this has been severe, the groove may be sufficiently deep for the nail to separate and it does not always grow back again, being replaced by an area of scarring.

Pterygium is a more unusual presentation: it is particularly characteristic of vasomotor ischaemia (Edwards, 1948). Spreading of the lunula, which leads to apparent leukonychia, is likewise an occasional manifestation of ischaemia.

Many peripheral arterial diseases (diabetes mellitus, Buerger's disease, etc.) can produce such changes, but they are particularly common in Raynaud's disease. This peripheral vascular disease of unknown aetiology involves principally the fingers, and is characterised by spasm with blanching, followed by a phase of hyperaemia. Raynaud's syndrome may occur as a complication of several conditions, ranging from the collagen diseases to brachial plexus compression. As in progressive scleroderma, sclerodactyly often begins with repeated attacks of Raynaud's phenomenon, eventually producing atrophic spindling of the fingers accompanied by considerable nail damage (Fig. 9.1). This is close to occupational acro-osteolysis caused by polymerisation of vinyl chloride.

Arteriography of the hand is almost normal in Raynaud's disease, and abnormality of a digital artery in a case of Raynaud's phenomenon should raise suspicion of some general disease, most commonly scleroderma.

COLLAGEN DISEASES

In systemic lupus erythematosus and dermatomyositis, erythematous bands are found on the dorsal aspect of the fingers and around the nails, associated with telangiectasia. These are of considerable diagnostic importance in some collagen diseases and are present in the CRST syndrome tetrad (cutaneous calcification, Raynaud's phenomenon, sclerodactyly, telangiec-

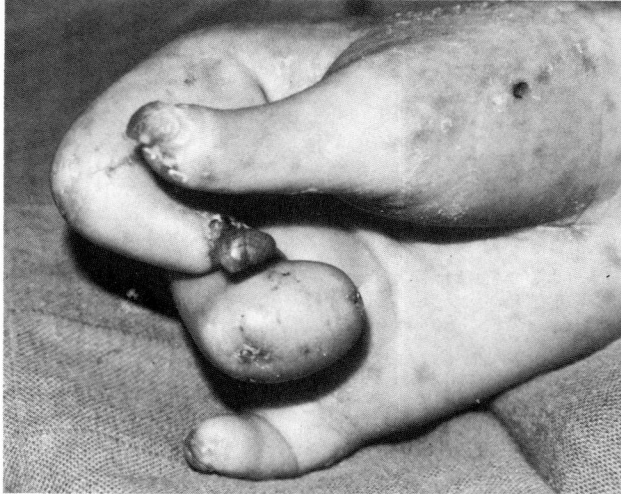
Figure 9.1
Acrosclerosis.

tasia). Such changes are almost pathognomonic of the collagen diseases, when they are found in the proximal nail fold (Plate 50a). They may take the form of linear telangiectasia within the cuticle sometimes with extravasation of blood (Plate 50b). Equally, they may be found in the tissues surrounding the nail.

Because of their oblique course, the capillaries of the proximal nail fold have been subject to considerable microscopic study and tentative diagnoses have been made on their appearance in certain pathological states.

Samitz has described other skin changes in dermatomyositis. The skin is often thickened, hyperkeratinised, rugose and irregular. These changes are not pathognomonic as they occur also in scleroderma, but they tend to parallel the severity of the disease.

GASTRO-INTESTINAL DISORDERS

Terry's cirrhotic white nail is a pseudo-macrolunular type of apparent leukonychia. Some cirrhoses produce flat, smooth, soft nails; others produce koilonychia (haemochromatosis), and some finger clubbing (chronic active hepatitis). Gastro-intestinal diseases account for 3.5 per cent of nail clubbing.

In the Cronkhite–Canada syndrome (adenomatous polyposis, with muco-cutaneous hyperpigmentation), the nails atrophy and may fall off.

YELLOW NAIL SYNDROME OF SAMMAN AND WHITE

The 'yellow nail syndrome' of Samman and White (1964) (Plate 51) in early descriptions associates the nail changes with lymphoedema. Subsequently the picture has been completed by the description of pulmonary effusions. Dilley et al., (1968) however, noticed in patients with normal nails the coexistence of

effusions and primary lymphoedema. A situation similar to, but not identical with, the 'yellow nail syndrome'.

Samman's nail is characterised by late onset (middle age), greenish discoloration, thickening and hardening of the nail and increased curvature. The lunula disappears and the cuticle is lost, chronic paronychia being common. Onycholysis is occasionally seen and spontaneous shedding also occurs. The growth of the nail is always greatly lessened.

INFECTIOUS DISEASES

Beau's lines in this context have been discussed earlier (Chapter 4). In subacute bacterial endocarditis the Osler's sign is of considerable diagnostic value. It is usually found in the pulp, although occasionally around or under the nail. It is a small, red, sensitive nodule, which develops over two to three days, or, in some cases, even hours. Some are so unobtrusive as to require systematic search. This may be accomplished by exerting pressure systematically over the area. Osler's false whitlow is found in acute and subacute bacterial endocarditis and is a pathognomonic sign. Finger clubbing is generally slight. Splinter haemorrhages in the distal half of the nail are attributed to bacterial emboli.

SARCOIDOSIS

Lesions of the nail may accompany the rare involvement of the terminal phalanges. The nail plate may become atrophied, cracked, deformed or thickened, convex and yellowish. Eventually the nail bed becomes reddened. Painful paronychia and splinter haemorrhages may occur (Fig. 9.2) (Sablet, 1971).

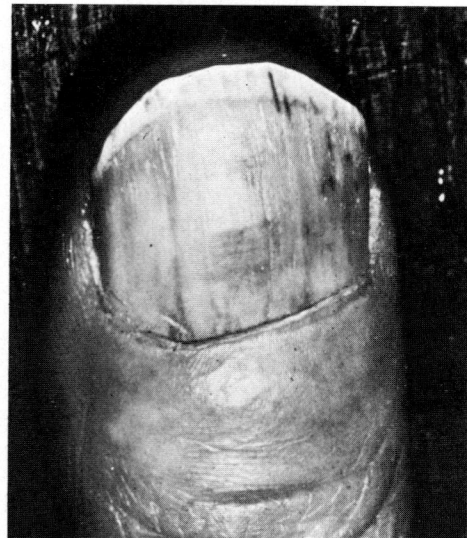
Figure 9.2
Sarcoidosis.

MULTICENTRIC RETICULOHISTIOCYTOSIS

This is characterised by a proliferation of histiocytes containing lipoid and giant cells. The hands, and in particular the fingers are a common site for the yellow or red–brown nodules, clustered around the nails like coral beads (Barrow, 1967). The nails atrophy, break and desquamate with longitudinal ridging. The destruction of the nail and distal phalangeal joint shortens the terminal segment, which resembles a raquet thumb.

LETTERER-SIWE DISEASE

The nail is rarely affected in this disease. Bender and Holzmann (1958), Kahn (1969) and Civatte *et al.* (1974) each report single cases with paronychia.

FOLLICULAR MUCINOSIS

In generalised follicular mucinosis, Lapierre *et al.* (1972) observed identical changes in the hands and feet. The affected nails were brittle, thickened and striated. Histologically, there was a dense histiolymphocytic cell infiltrate in the nail bed separating the basal cell layer and the stratum mucosum, similar to that surrounding the hair follicle. The entire thickness of the nail was hollowed out by cavities running parallel to the surface and filled by amorphous material.

HODGKIN'S DISEASE

Shahani and Blackburn (1973) described nail changes in Hodgkin's disease, which they associated with poor prognosis.

Their patients presented with one to three transverse striae of leukonychia on each nail. In one of them a brown coloration of the distal part contrasted with the paler proximal segment.

ENDOCRINE DISORDERS

In myxoedema the nails become dry, brittle and ridged, the hands cold and the fingers swollen. There may be koilonychia in hypothyroidism and in acromegaly, and onycholysis is seen equally in hypo- and hyperthyroidism. Clubbing, associated with exophthalmus and pretibial myxoedema is seen in Diamond's syndrome. In hyperparathyroidism, the nails are dry, brittle and grooved, chronic paronychia may be seen (Fig. 9.3).

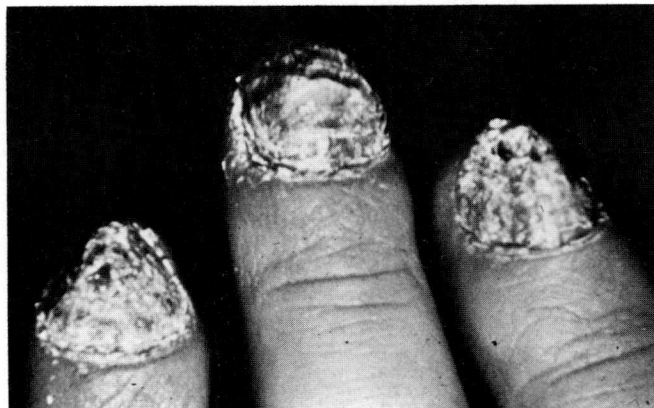

Figure 9.3
Candida granuloma in hypoparathyroidism.

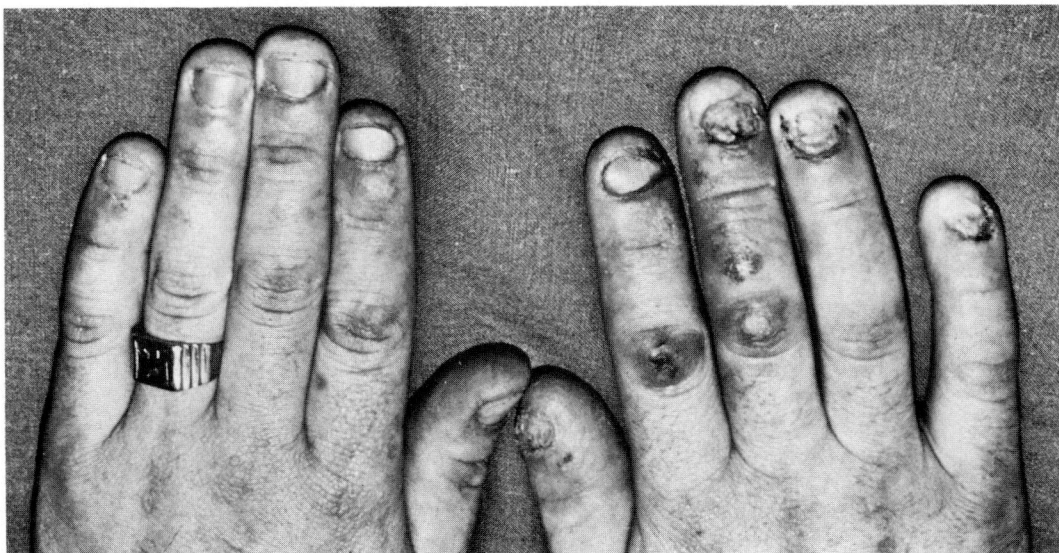

Figure 9.4
Nail dystrophy in unilateral syringomyelia (Mme C. Beylot's collection).

DISEASES OF THE NERVOUS SYSTEM

Traumatic and experimental lesions of the central nervous system have confirmed the existence of a hypothalamic centre, which determines the characteristics of the nail (consistency, elasticity and growth). Cases with a similar involvement of isolated digits have led to the supposition that there must be peripheral centres in addition. Changes parallel the hypo- or hyper-function of the hypothalamus.

In Morgagni–Stewart–Morel syndrome, nails become extremely thickened and hard, appearing as a dull cornified mass. Their rapid uncontrolled growth produces a type of onychogryphosis. In syringomyelia, dystrophic nail changes accompany reduction or cessation of growth (Fig. 9.4).

Onychomadesis is found classically in tabes, cerebral confusional states and epilepsy (where fits may be marked by Beau's lines). Nail growth is occasionally accelerated in Parkinson's disease, while in hemiplegia, the nails of the affected limbs grow more slowly, as in disseminated sclerosis, in which a variety of nail dystrophies also occur.

Tuberous sclerosis (epiloia) presents with Koenen's tumours. In Lesch–Nyhan's syndrome in addition to the nervous and biological symptomatology there is also mutilation of the fingers. Finally onychodystrophies of various forms are frequently encountered in injuries of the peripheral nervous system.

BIBLIOGRAPHY

[1] BARAN R., TEMIME P., COUNILLON-LEULIER J. — Trophic nail problems of circulatory origin. International Congress Series n°. 289. *Proceedings of XIV International Congress of Dermatology*, Edited by E. Flarer et F. Serri, *Padua-Venice, 22–27 May 1972*. Amsterdam, Excerpta Medica.

[2] BARROW M.-V. — The nail multicentric reticulohistiocytosis. *Arch. Derm. (Chic.)*, 1967, *95*, 200–201.

[3] BRAVERMAN I.-M. — *Skin signs of systemic disease*. Philadelphia, London, Toronto, W. B. Saunders, 1970, p. 155–156.

[4] DILLEY J.-J., KIERLAND R.-R., RANDALL R.-V. — Primary lymphoedema associated with yellow nail and pleural effusions. *JAMA*, 1968, *204*, 670–673.

[5] EDWARDS E.-A. — Nail changes in functional and organic arterial disease. *New Engl. J. Med.*, 1948, *239*, 362–365.

[6] SABLET M. de. — Sarcoïdose avec maladie de Perthes-Jungling et onyxis. *Bull. Soc. franç. Derm. Syph.*, 1971, *78*, 289–291.

[7] SAMMAN P., STRICKLAND B. — Abnormalities of the finger nails associated with impaired peripheral blood supply. *Brit. J. Derm.*, 1962, *74*, 166–173.

[8] SAMMAN P., WHITE W. — The yellow nail syndrome. *Brit. J. Derm.*, 1964, *76*, 153–157.

[9] TERRY R. — Red half moons in cardiac failure. *Lancet*, 1954, *2*, 842–844.

10. NAIL CHANGES IN PRINCIPAL GENETIC DISEASES

R. Baran

The classification by J. M. Robert (1977) of genetic diseases in general is adapted here to disorders of the nail, under three headings:

1. Genetic malformations with primary involvement of the nail (Tables 10.1 and 10.2).
2. Genetic abnormalities involving tissues or groups of tissues, in which the nail involvement is incidental (Table 10.3).
3. Genetic disorders of metabolism with nail changes (Table 10.4).

TABLE 10.1 Malformative hereditary diseases: chromosome anomalies

Disease	Age of onset	Nails	Clinical elements of corresponding syndrome
Monosomia 4 p	constitutional	no lunulae, narrow and convex	gross mental deficiency; microcephaly; facial dysmorphia (hypertelorism, epicanthus, wide nose bridge, etc.); cardiac malformation
Trisomia 7 q	constitutional	abnormally convex	gross mental deficiency; dysmorphia of face and skull; hypotonia
Trisomia 8	constitutional	hypoplastic; may not be present at birth	moderate mental deficiency; facial dysmorphia; thickened, everted lower lip; bone and joint anomalies (stiffness); deep palm and plantar creases
Trisomia 8 p	constitutional	dysplasia, especially in toe-nails	mental deficiency; sometimes dysmorphia; minor abnormalities of hands and feet
Trisomia 9 p	constitutional	dysplasia and sometimes clawshape, especially in the toes	variable mental defect; brachycephaly; dysmorphias: huge nose, peculiar grimace
Monosomia 9 p	constitutional	wide, square and increased convexity	variable mental defect; trigonocephaly; slanting eyes; long upper lip
Trisomia 13	constitutional	narrow and very convex	microphthalmia; hexadactyly in two-thirds of cases; multiple deformities (especially of the heart)
Trisomy 18	constitutional	hypoplastic	abnormalities of hands and feet; multiple visceral deformities
Turner's syndrome (45XO, Berger, 1972) Bonnevie–Ullrich's syndrome	constitutional	deformed, very convex, hypoplastic; sometimes missing	Bonnevie–Ullrich's syndrome: transient lymphoedema of fingers and toes common characteristics of both syndromes: short stature, ovarian dysgenesis and visceral deformities

TABLE 10.2 Hereditary deformities with normal karyotype

Disease	Age when affected	Nails	Symptomatology	Transmission
(a) *Simple genetic mutation*				
Congenital pterygium	birth			?
Ventral pterygium	birth	oblique fractures	aberrant hyponychium; can be painful	?
Leukonychia (Baran and Achten, 1969)	birth			AD
Koilonychia	birth			AD
Anonychia	birth			AD or R
(b) *Mutation of a pleiotropic gene*				
Osteo–onychodysplasia (nail patella syndrome) (Fig. 10.1) (Gibbs et al., 1964 Manigaud et al., 1971)	birth	constant pathognomonic, bilateral and symmetrical abnormalities, especially of the hand; usually attacks thumb and lesions, become less serious towards little finger; nail is missing or hypoplastic; shorter and narrower than usual (sometimes only ⅓ of usual width); is also fragile, brittle; may have a longitudinal groove (central ptergyium); lunula often missing or shaped like inverted V	patellae are missing or hypoplastic, dysplasia of the radial head, iliac horns, nephropathies, linkage with gene responsible for ABO blood group	AD
Focal dermal hypoplasia or Goltz's syndrome	birth	thin, narrow, dystrophic; koilonychia; lunula missing; frequent anonychia	pseudo-bullae; telangiectasia; linear pigmentation; abnormalities of bones and eyes	AR d
Apert's syndrome (acrocephalo-syndactyly)	birth	atrophy of nail	cranio-stenosis: prominent eyes; syndactyly	AD
Ellis van Creveld's syndrome (chondroectodermal dysplasia)	birth	small, thin, short, striated, atrophied; ocasionally koilonychia	dwarfed, polydactyly, abnormalities of heart, skeleton and teeth	AR

(continued on next page)

TABLE 10.2 *(continued)*

Disease	Age when affected	Nails	Symptomatology	Transmission
Rubinstein–Taybi's syndrome (Rubinstein and Taybi, 1963)	birth	wide thumbs and toes	dysmorphia of face, retarded mental and physical growth, antimongoloid obliquity of eyes, beaked nose, low set ears	AR d
Glossopalatine ankylosis	birth	anonychia	cleft palate is common; tongue fixed to palate or edge of alveoli; reduced number of teeth; variety of hand and foot abnormalities	?
Popliteal pterygium syndrome (McKusick, 1971)	birth	onychodysplasia	cleft palate; hypoplasia or agenesis of fingers; syndactyly; genital abnormalities	AD or R

Notes:

 A = autosomal
 D = dominant
 R = recessive
 d = form of transmission still under discussion
 S = sex-linked transmission

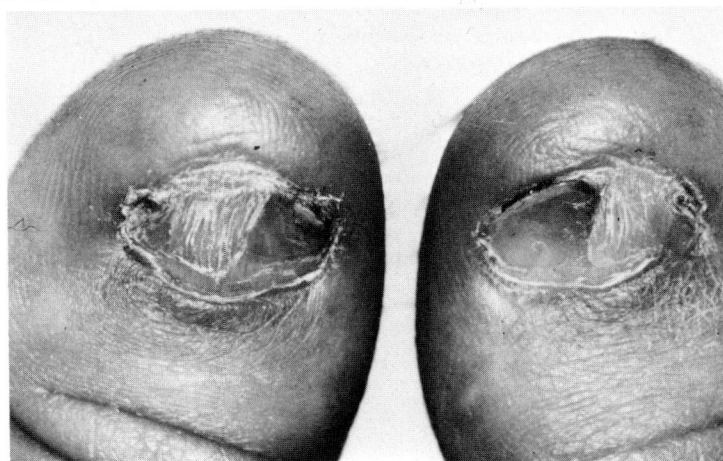

Figure 10.1
Nail-patella syndrome.

TABLE 10.3 Hereditary diseases of the tissues

Disease	Age when affected	Nails	Symptomatology	Transmission
(a) *Neuro-ectodermoses* Epiloia	before age of 5 years	Koenen's tumours; subungual fibromas	sebaceous adenomas of face; visceral abnormalities	AD
(b) *Ecto-dermoses* Ichthyosis vulgaris	1–4 years	hyperplasia	elbows and knees affected by diffuse desquamation without erythema	AD

(continued on next page)

TABLE 10.3   (*continued*)

Disease	Age when affected	Nails	Symptomatology	Transmission
Lamellar ichtyosis (non bullous congenital ichthyoform erythoderma) (Fig. 10.2)	birth	thickened or striated nails, accelerated growth	general dry erythroderma, and hyperkeratosis, especially at joints	AR
Palmoplantar Keratodermia, Meleda type	neonatal	koilonychia, sub-ungual hyper-keratosis, nails often short	palmo-plantar keratotic erythema	AR
Palmoplantar Kerarodermia Thost–Unna's type	first 6 months	hyperplasia	palmo-plantar hyperkeratosis with hyperhidrosis	AD
Pachyonychia congenita (Joseph, 1964)	birth	pachyonychia with hypertrophy of nail bed	vesicles appear in early stages; later produce keratoses, leucokeratoses of mucous membranes, cataract	AD
Pityriasis rubra pilaris	childhood to adulthood	hyperplasia, opaque with longitudinal ridges and subungual hyperkeratosis	eruption of follicular, hyper-keratotic papulae	AD
Darier's disease	8–16 years	subungual, V-shaped keratoses, red and white longitudinal bands	pigmented, keratotic papulae, especially on trunk	AD
Benign familial acanthosis nigricans	late childhood to puberty	thickened, dull and brittle	thickened, papillomatous pigmented skin at major joints, neck and trunk	AD
Epidermolysis bullosa (hyperplastic) (Fig. 10.3)	babyhood to puberty	hyperplasia	post-traumatic bullae with occasional scarring	AD
Epidermolysis bullosa (hyperplastic) (Fig. 10.4)	birth and first week	rudimentary nails	widespread bullous elevation of skin and mucosa with scarring	AR
Bart–Gorlin–Anderson's syndrome or localised absence of skin with blistering and and dystrophy (Bart *et al.*, 1966)	birth	congenital anonychia or secondary nail fall, or onychogryphosis, grey–yellow colour	skin hypoplasia in lower limbs; bullae on skin and/or mucous membranes, which do not leave scars	AD

(continued on next page)

TABLE 10.3 (*continued*)

Disease	Age when affected	Nails	Symptomatology	Transmission
(c) *Ecto-endodermoses*				
Peutz–Jeghers–Touraine's syndrome	infancy	rare diffuse or longitudinal melanin pigmentation; occasionally clubbing	pigmentation of lips and mucous membranes; polyposis of intestines; ovarian tumours	AD
(d) *Ecto-mesodermoses*				
sclerotylosis Huriez's syndrome (Huriez, 1969)	birth	varying from simple grooves to complete aplasia	frequent squamous cell transformation in hands and feet; linkage with gene responsible for blood group MN	AD
Pachydermo-periostosis (Hambrick and Carter, 1966; Huriez *et al.*, 1962)	before age of 20 years	clubbing	pachydermia of hands and feet and face, periosteal thickening of hands and feet and irregular pachydermia of scalp	AD
Myotonic dysplasia, Steinert's type	between ages 20 and 30 years	nail atrophy	atrophy of hands and feet, hair; cataract; myotonia	AD
Werner's syndrome	between ages 20 and 30 years	nail atrophy	severe scleroderma with atrophy of hands and feet; cataract	AR d
Variot and Cailliau's infantile gerodermia	6 months to 1 year	nail atrophy	diffuse atrophy with skin and premature senescence	AR d
Progeria (Hutchinson–Gilford's syndrome)	first months	delicate and brittle	dwarfism and senile appearance	AR d
Acrogeria (Gottron's syndrome)	birth	atrophic or thickened onychogryphosis–clubbing or koilonychia	skin atrophy, especially in hands and feet	AR d
Congenital poikilodermia (Rothmund–Thomson's syndrome)	3 months to 2 years	sometimes healthy, can be hypoplastic and dystrophic	poikilodermia; anomalies of skeleton, and teeth; cataract; hypogonadism; slow growth	AR d

(*continued on next page*)

TABLE 10.3 *(continued)*

Disease	Age when affected	Nails	Symptomatology	Transmission
Ectodermal dysplasia with anhydrosis (Weech or Christ–Siemens syndrome)	birth	delicate or brittle striated nails, koilonychia	hypotrichosis, edentulous problems of temperature control due to reduction in, or absence of, sweat glands, absent sebaceous secretion, sometimes syndactyly	RS
Hydrotic ectodermal dysplasia (Clouston's syndrome)	childhood	thickened, striated, dyschromic; growth is slow; more rarely, short, delicate, brittle with subungual hyperkeratosis onycholysis	palmo-plantar keratodermia, which may become malignant; hair is sparse, brittle, or absent	AD
Incontinentia pigmentia (Pinol Aguade *et al.*, 1973)	birth	delicate, flat with koilonychia and fibroma in the shape of a clove of garlic	successive and periodic bullae, verrucoid stage and pigmentation; anomalies of eyes, heart, teeth and brain	RS
	at 15 years	Subungual keratotic tumours, which clear up spontaneously		
Bart and Pumphrey's syndrome	early childhood	total leukonychia with koilonychia	deafness; palmo-plantar keratoses knuckle pads	AD
Cleido-cranial dysostoses	birth	bulging, sometimes brittle; generally in bad condition	late closure of fontanelles; brachycephaly; aplasia of clavicles; small stature; excess teeth	AD
Dyskeratosis congenita (Barriere, 1970) (Fig. 10.5)	5–12 years	short, friable, recurring paronychia, nail frequently falls off	leucoplasia of tongue, palmo-plantar hyperkeratosis; marrow aplasia; poikilodermia	AR
Pycnodysostosis	birth	irregular and cracked	small in stature; huge skull with frontal and occipital bosses; anterior fontanelle; closure delayed	AR

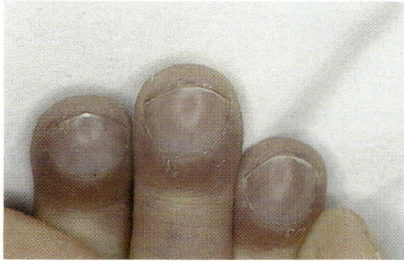
49

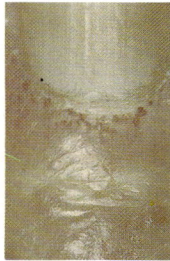
50 a

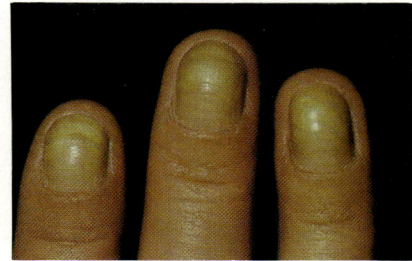
51

50 b

Plate 49   Hippocratism with cyanosis

Plate 50a Dermatomyositis. Telangectiasites of the cuticle ; extravasation of the blood

Plate 50b Dermatomyositis. Modification of the cuticle with extravasation of the blood

Plate 51   Sammon and White's xanthonychic syndrome (Grupper's collection)

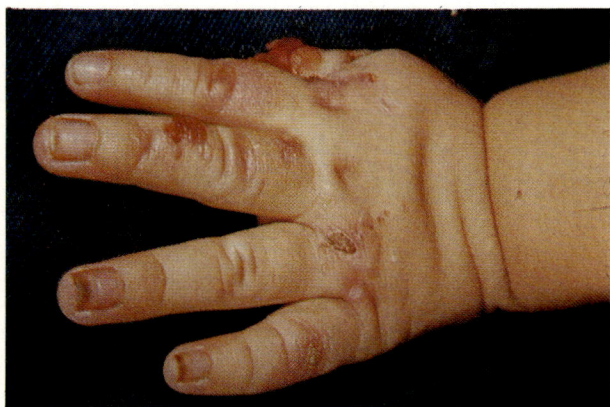

52 a

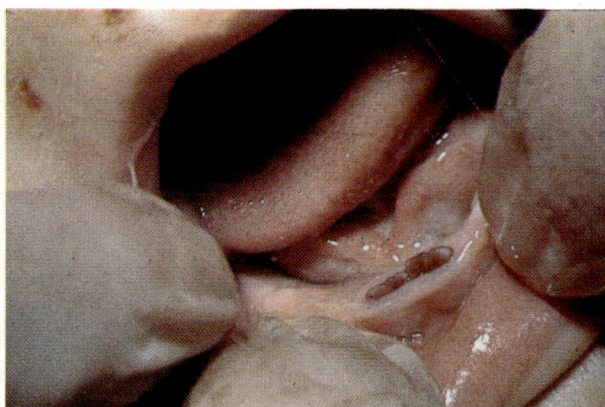

52 b

Plate 52a Congenital porphyria, Gunther's type.

Plate 52b Congenital porphyria, Gunther's type. Erythrodonia (Mascaro's collection)

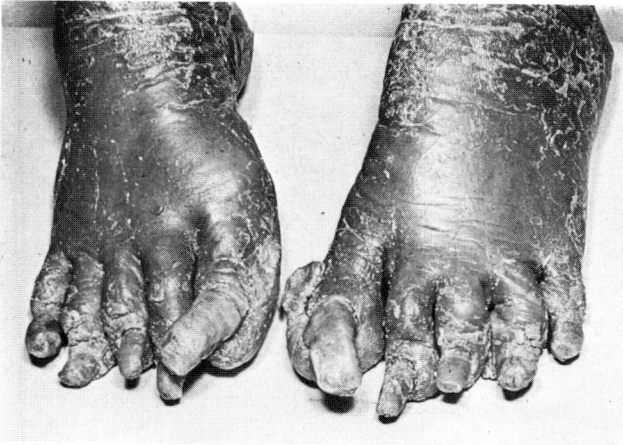

Figure 10.2
Congenital ichthyoform erythroderma (Puissant's collection).

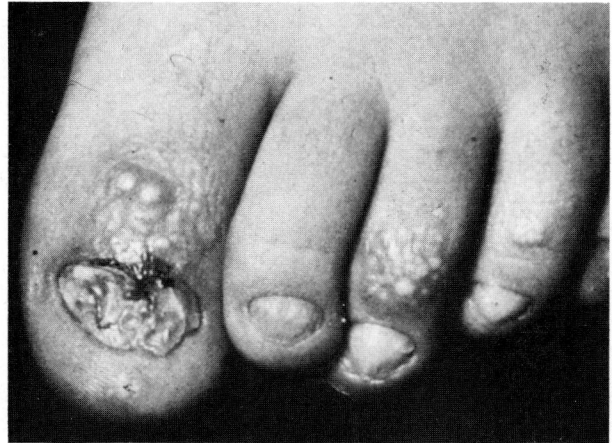

Figure 10.3
Hyperplastic bullous epidermolysis (Laugier's collection).

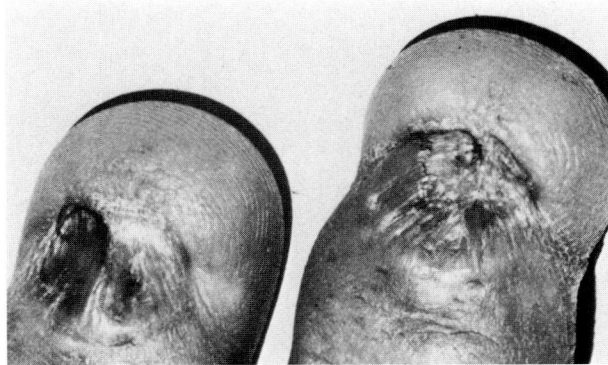

Figure 10.4
Hypoplastic bullous epidermolysis.

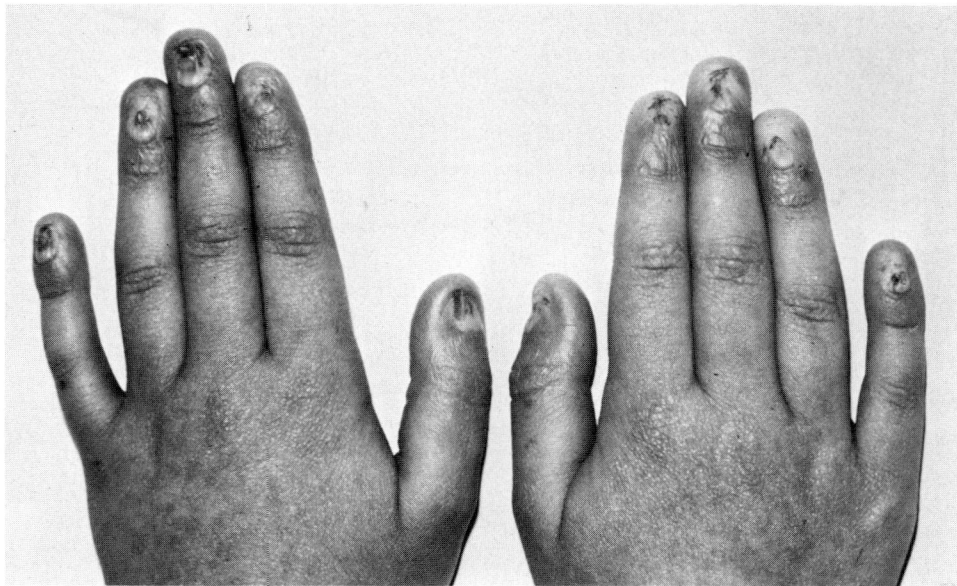

Figure 10.5
Dyskeratosis congenita (Barrière's collection).

TABLE 10.4 Genetic disorders of metabolism

Disease	Age when affected	Nails	Symptomatology	Transmission
Acrodermatitis enteropathica (Fig. 10.6)	end of first year	paronychia, candidiasis	eczematous lesions round orifices and psoriatic ones on elbows, knees, hands and feet; alopecia; malabsorption syndrome, which can be ameliorated by the use of zinc	AR
Shelley and Rawnsley's syndrome (Shelley and Rawnsley, 1965)		brittle	alopecia; problems with metabolism of argininosuccinic acid in urine	AR d
Childhood Haemochromatosis		total leukonychia, pseudomacrolunula (Terry's type); grey or brown nails; koilonychia (50%)	problem with iron metabolism; melanodermia, cirrhosis, diabetes	AD d
Lesch–Nyhan's syndrome (Silvers, 1972) (Fig. 10.7)	first months	auto-multilation of hands and feet	diffuse neurological disorders, oligophrenia, hyperuricaemia	RS
Ochronosis alkaptonuria	birth	grey–blue (when patient is fully-grown)	urine turns black when exposed to air; bluish skin colour due to transparency of skin covering cartilages and superficial tendons, which are impregnated with ochronotic pigment; osteoarthritis when patient is aged 20–30	AR
Congenital porphyria, Gunther's type (Plate 52)	before one year	brown colouring, possible mutilation of hands and feet	Vacciniform hydroa, hypertrichosis, erythrodontia, haemolytic anaemia	AR
Porphyria cutanea tarda	adult	diffuse pigmentation or pigmentation in bands; no lunula; subungual bullae; onychodysmorphias, especially koilonychia	bullae on parts exposed to sun; pigmentation and fragile skin	AR
Erythropoietic proto-porphyria.	early childhood	no lunula; nails are opaque, bluish, or dark brownish	skin burns immediately in sun; urticaria; scars; hyperkeratoses; pigmentation; normal urine	AR
Wilson's disease	childhood or young adult	'Azure half moons'	spasmodic rigidity with tremours; bronzed sclero-corneal ring; psychic problems; nodular hepatitis with cirrhoses; problems with copper metabolism	AR

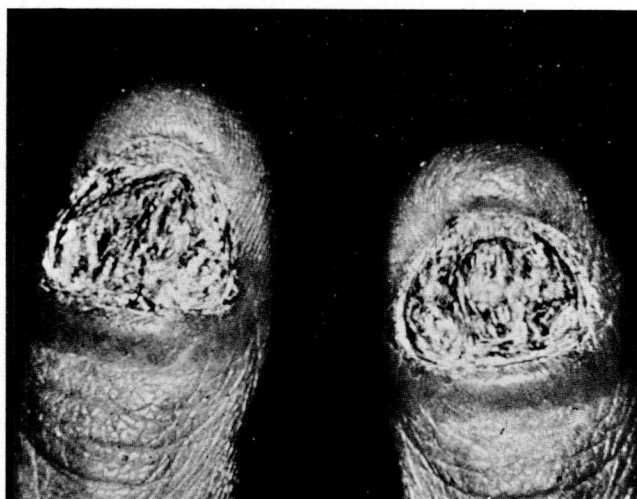

Figure 10.6
Acrodermatitis enteropathica (Grupper's collection).

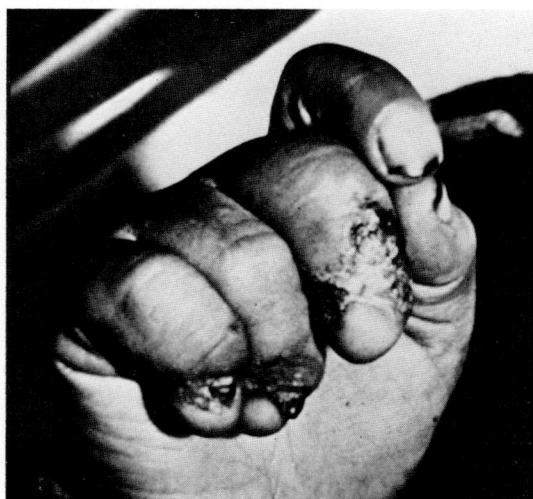

Figure 10.7
Lesch-Nyhan's syndrome.

BIBLIOGRAPHY

[1] BARAN R., ACHTEN G. — Les associations congénitales de koïlonychie et de leuconychie totale. *Arch. belges Derm. Syph.*, 1969, *25*, 13–29.

[2] BARRIERE H. — Dyskératose congénitale avec thrombopénie. Ses relations avec l'anémie de Fanconi. *Sem. Hôp. Paris*, 1970, *46*, 3083–3087.

[3] BART B.-J., GORLIN R.-J., ANDERSON V.-E., LYNCH F.-W. — Congenital localized absence of skin and associated abnormalities ressembling epidermolysis bullosa. *Arch. Derm. (Chic.)*, 1966, *93*, 296–304.

[4] BART R.-S., PUMPHREY R.-E. — Knuckle pads, leukonychia and deafness. *New Engl. J. Med.*, 1967, *276*, 202–206.

[5] BEHAR A., RACHMILEWITZ E. — Ellis-van Creveld syndrome. *Arch. intern. med.*, 1964, *113*, 606–611.

[6] BERGER R. — Eléments de génétique médicale. *Encycl. méd. chir.*, Pédiatrie, 1972, 4002 T 10, T 50.

[7] GIBBS R., BERCZELLER P.-H., HYMAN A.-B. — Nail-patella-elbow syndrome. *Arch. Derm. (Chic.)*, 1964, *89*, 196–199.

[8] HAMBRICK G.-W., CARTER M. — Pachydermoperiostosis. *Arch. Derm. (Chic.)*, 1966, *94*, 594–608.

[9] HURIEZ C., DEMINATI M., AGACHE P., DELMAS-MARSALET Y., MENNECER M. — Génodermatose scléro-atrophiante et kératodermique des extrémités. *Arch. derm. syph. (Paris)*, 1969, *96*, 135–146.

[10] HURIEZ C., FRANÇOIS P., AGACHE P. — Pachyder-mopériostose. *Arch. derm. syph.* (Paris), 1962, *89*, 372–381.

[11] JOHNSON C.-F. — Broad thumbs and broad great toes with facial abnormalities and mental retardation. *Pediatrics*, 1966, *68*, 942–951.

[12] JOSEPH H.-L. — Pachyonychia congenita. *Arch. Derm. (Chic.)*, 1964, *90*, 594–603.

[13] MANIGAUD G., AUZEPY P., PAILLAS J., COHEN de LARA A., DEPARIS M. — Ostéo-onycho-dysplasie avec néphropathie. *Sem. Hôp. Paris*, 1971, *47*, 2949–2956.

[14] McKUSICK V.-A. — *Inheritance in man*. Baltimore, The Johns Hopkins Press, 3ᵉ edit., 1971.

[15] PINOL AGUADE J., MASCARO J.-M., HERRERO C., CASTEL T. — Tumeurs sous-unguéales dyskératosiques douloureuses et spontanément résolutives. Ses rapports avec l'Incontinentia Pigmenti. *Arch. derm. syph. (Paris)*, 1973, *100*, 159–168.

[16] ROBERT J.-M., PLANCHU H., GIRAUD F., MATEI J.-F. — In: *Génétique et cytogénétique cliniques*. Paris, Flammorion Médecine Sciences, 1977.

[17] RUBINSTEIN J.-H., TAYBI H. — Broad thumbs and facial abnormalities. *Amer. J. Dis. Child.*, 1963, *105*, 588.

[18] SHELLEY W.-B., RAWNSLEY H.-M. — Aminogenic alopecia. *Lancet*, 1965, *2*, 1327–1328.

[19] SILVERS D.-N. — Lesch-Nyhan disease: An X linked recessive inborn error of purine metabolism. *Arch. Derm.*, 1972, *105*, 933–934.

[20] TOURAINE A. — *L'hérédité en médecine*. Paris, Masson, 1955.

11. MELANOTIC TUMOURS IN THE NAIL AREA

P. Banzet, R. Mouly and C. Dufourmentel

The diagnosis of melanotic tumours in the nail area presents several difficulties and it is with these problems that the following pages are mainly concerned. Since the nail area is exposed to everyday injury, it is easy to dismiss lesions in this region as benign with the result that the detection of malignant lesions is frequently delayed with serious consequences. Awareness of the possibility of malignant change in this area should minimise such delays but accurate diagnosis, even by experts, can be very difficult; one of the problems being that a number of malignant melanomas lack pigmentation. In all cases of doubt, a biopsy should be obtained for histological examination.

Traditionally, malignant tumours of the nail area are divided into two groups: those located around the nail and those under the nail.

PERIUNGUAL LESIONS

Periungual lesions should not be difficult to diagnose even though they may have a variable presentation. The most typical appearance is a pigmented tumour with exuberant granulation starting in the nail fold but making its way steadily towards the subungual area. Its colour and tendency to bleed readily should raise the alarm. Difficulties arise with the amelanotic lesions, which closely resemble a chronic paronychial infection. Diagnosis may not be possible on first examination but the chronicity of the lesion should cause concern. However, the patient himself may be misled by the indolence of the lesion and delay seeking medical advice. Early biopsy is essential in all doubtful cases.

SUBUNGUAL MALIGNANT MELANOMAS

Diagnosis of subungual malignant melanomas may be more difficult: clinical differentiation between benign and malignant lesions can be virtually impossible, yet benign melanomas in this region are very common, and other conditions, such as subungual haematomas, can mimic melanomas very closely, although usually a clear history of trauma can be obtained, which together with the acute and painful onset should resolve doubt. However, chronic haematomas may arise, usually in the toes rather than in the fingers, which can cause confusion. This may be resolved by avulsion of the nail, which removes the haematoma as well as exposing a healthy bed and matrix.

Other subungual lesions, which can be deceptive, include the glomus tumour, which appears as a small blue lesion and is not invariably painful; subungual exostoses may push up and deform the nail as well as causing an inflammatory granulation, but here an X-ray is diagnostic; some fungal infections are deceptive because of their colour and chronicity, but it is unusual for the nail to become deformed in such lesions.

Typical subungual melanomas present as a pigmented band under the nail and extend the whole length of the nail. Traditionally it is maintained that, if the nail groove is affected (Hutchinson's sign), malignancy is present. However, the validity of this assertion is questionable. Tradition also has it that an inflammatory reaction around the nail only occurs if the lesion is malignant, but this also is not always the case. Many consider that the only unequivocal sign of malignancy is destruction of the nail by the underlying, fungating tumour and, while this indeed indicates malignancy, it does so at a stage which may be too late for effective treatment. Even this appearance cannot be considered as absolute proof so that biopsy is still essential. Some doctors regard all subungual pigmented lesions as potentially malignant melanomas, which may be taking things too far, but, when in doubt, the only safe solution is to carry out a biopsy. If the matrix is intact, there is less chance of the melanoma being malignant. If pigment involves the matrix, the lesion should be resected *en bloc*, including part of the plate, the underlying bed and the matrix. Histological examination will usually establish the diagnosis, but even the pathologist has difficulty in some cases. The earlier the excision is done, the better the prospect of permanent cure.

All reports relating to subungual and periungual malignant melanomas indicate long delays between the appearance of the lesions and diagnosis. This is particularly disappointing considering the relatively good remission, which can be achieved in comparison with other types of malignant melanomas. On the other hand, a patient with a granulating tumour, which has burst through the nail and spread to the lymph nodes, has a very poor prognosis and such a condition can develop relatively rapidly. Such cases are rather unusual and in the majority of cases there is an opportunity for early action, which can prevent metastases.

TREATMENT

Although this book is concerned with sugery of the hand, it must be emphasised that the role of surgery in this tumour is more limited than in the past and major mutilating amputations are no longer practised now that it is becoming clear that a different approach is required to control the spread of this very immuno-dependent tumour. Surgery is still necessary to confirm the diagnosis and to remove the primary tumour but other biological processes are required to re-establish equilibrium, control the spread of metastases and bring about remission. The extent of surgery is greatly reduced, if treatment is carried out in collaboration with chemotherapists and immunologists. To lose a finger is acceptable, but to lose a thumb is much less so, particularly when the tumour is small. The value of any surgical technique alone is in doubt and a multidisciplinary approach is indicated. Many surgeons have bitter memories of unsuccessful surgical treatment in the past and recognise the need for a change of approach. The hand surgeon may find himself confined to a minor role in a well co-ordinated regime, which allows early diagnosis and a better chance of success without loss of function of the hand.

12. GLOMUS TUMOURS

H. Bureau, J. P. Jouglard, A. Thion, H. Tramier and M. Pierre

Known also as painful subcutaneous nodule (W. Wood, 1812, Chandelux, 1882), perithelioma (Muller, 1901), neuromyoarterial glomangioma (Masson, 1924) and haemangiopericytoma, the glomus tumour is a small benign lesion, most commonly found in the skin or cellular subcutaneous tissue. It is rare, but by no means exceptional. It is usually single and most often found in the fingers. The chief symptom is pain, and treatment is complete surgical removal.

Direct arterio-venous shunts, excluding capillary networks, are found at several sites in the body. Some are anomalous, while others, the glomus bodies, are specific and precisely located. They are exclusive to warm-blooded vertebrates. In man, they are found mainly in the dermis of the extremities: the hand, the foot, the ears, the tip of the nose. They are most numerous in the nail bed (10–20 per cm²).

A glomus body is a tiny, oval organ, 30–300 microns long, composed of: a tortuous vessel uniting an arteriole and a venule; a nerve supply and a capsule.

The microscopic structure is specific: some of the preterminal cutaneous arterioles bifurcate into a branch, which ends in a capillary bed, and a thicker-walled branch, which after a short course of approximately 30 microns, joins a venule. At this site, the arteriole loses its internal elastic lamina and its wall thickens, due to the incorporation of smooth muscle cells between the endothelial and medial layers. In transverse section, these cells have the appearance of epithelioid cells. They assume globular shape on contraction, pushing the endothelium centrally, obliterating the lumen. When relaxed, the vessel wall expands outwards. The adventitia of the anastomotic vessel is replaced by a fibrous sheath supplied by myelinated sensory nerve fibres and unmyelinated motor fibres, some of which penetrate as far as the epithelioid cells. These structures are sheathed by capsule, which separates them from the subdermal tissues.

The glomus bodies lie parallel to the capillary reservoirs (E. R. and E. L. Clarke, 1934). They can contract rhythmically at a slower rate than that of the arterioles and asynchronously with them, also responding to stimulation in a different manner: for example, the arteriolar response to cold is to contract, while the glomus body dilates. These are, therefore, the regulators of the capillary circulation (the 'Peripheral heart' of Masson), which share in the balance between the cooling of blood in the extremities and the maintenance of central arterial blood pressure.

The local variations in blood pressure, produced by the activity of glomus bodies, influence the formation of interstitial fluid and regulate the tension of the pulp connective tissue, which is important for proper functioning of the tactile receptor organs of the dermis.

The complex neural apparatus of the glomus bodies gives them a special sensitivity and damage, or absence, leads to circulatory and trophic problems.

PATHOLOGICAL ANATOMY

The glomus tumour consists usually of a disordered mass of intertwining vessels, the arterial segment adjoining multiple venous segments, dominated by syncytial epithelioid cells and unmyelinated nerve fibres.

CLINICAL FEATURES

The pathognomonic feature of the glomus tumour is pain: spontaneous pain is rare, but common in tumours of long standing. Induced pain is a major sign. Any contact, in fact the slightest brush, for example, with a glove, face-cloth or clothing, or examination, causes the patient to wince. The intensity of the pain is extreme and its description depends on the extent of the patient's vocabulary (Leriche, 1936). It is of short duration and subsides completely between attacks. It tends to become progressively worse and the development of radiation is an inconstant feature (the 'hyperalgie hyperdiffusante' of Roger and Alliez). The psychological repercussions of this intense dominating pain have often led to treatment of such patients as if they were mentally ill.

The origin of this pain is the subject of debate. However, it appears to be the resultant of a radiating pain of spinal origin and local pain triggered by distension of the neural apparatus of the glomus body.

EXAMINATION

Tumours are always small, varying in size between a pinhead and a cherry stone, rarely larger. Under the nail, it appears as a small bluish area (Fig. 12.1). On further examination, the following may be found:
(a) changes in the local circulation – flushing, elevation or depression of the temperature
(b) trophic lesions – muscular atrophy, scalloping of bone evident radiologically
(c) neurological problems – reduced skin sensitivity, depressed reflexes, excessive sweating.

The typical site is the dermis of the nail bed, most commonly the ventral surfaces of the fingers and toes, and occasionally elsewhere in the body.

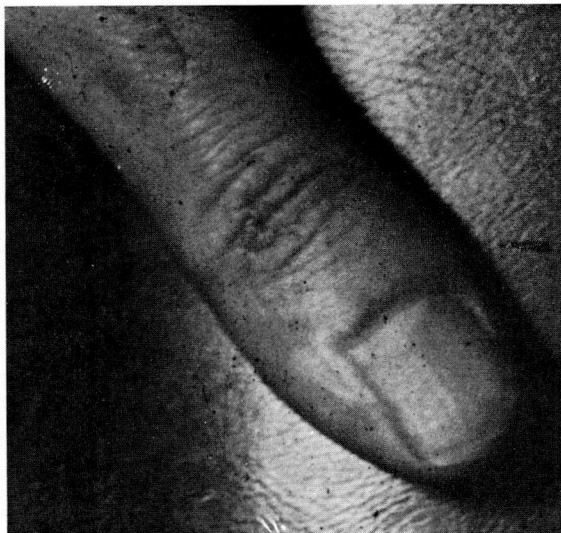

Figure 12.1
Glomus tumour on the side of the nail, which had already been operated upon and wrongly diagnosed.

Diagnosis may be difficult and is frequently delayed. Leriche described a patient who suffered from a tumour in the armpit of thirty years standing. In one of our thirteen personal cases, a glomus tumour in the pulp of a finger had undergone several incomplete excisions and percussion sessions.

CLINICAL TYPES

Painless variants exist that are not sensitive to stimulation. These are usually found in cases where there are multiple tumours, some of which are painful and others not, and in early lesions.

HISTOLOGICAL TYPES

Epithelioid forms with distinct columnar arrangement, predominantly angiomatous forms and predominantly neural forms, can be distinguished (Figs. 12.2 and 12.3).

CLASSIFICATION ACCORDING TO SITE

Glomus tumours may be found in many sites (muscles of the arm, uterus, and knee); the glomus body is not, therefore, exclusive to the skin. Some are found in bony lacunae in the terminal phalanx (three of our thirteen cases), but in these instances it is difficult to differentiate the primary tumour from a secondary extension of the tumour of the pulp.

TREATMENT

Iselin (1955, 1967) points out that the common description of the lesion as 'a painful subungual tumour' has resulted in many inadequate surgical procedures being directed to the bed of the

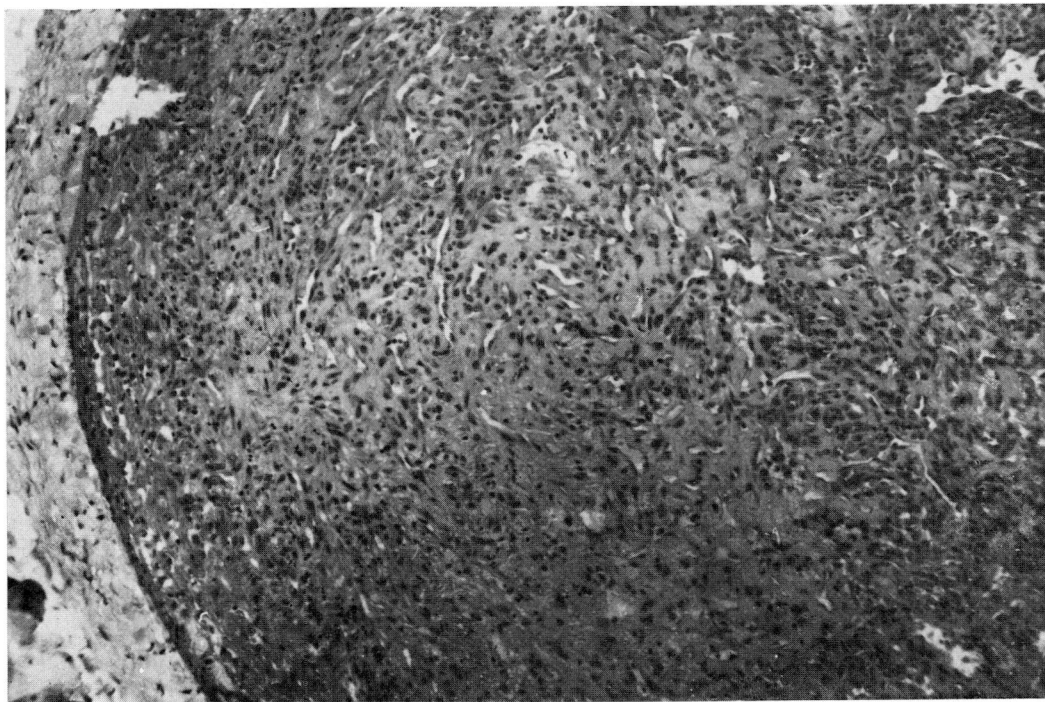

Figure 12.2
Glomus tumour, hyalinised form. A well-encapsulated lesion, consisting of glomus vessels and cells contained in a fibrous conjunctive stroma (Pr. Bonneau) (magnified × 100).

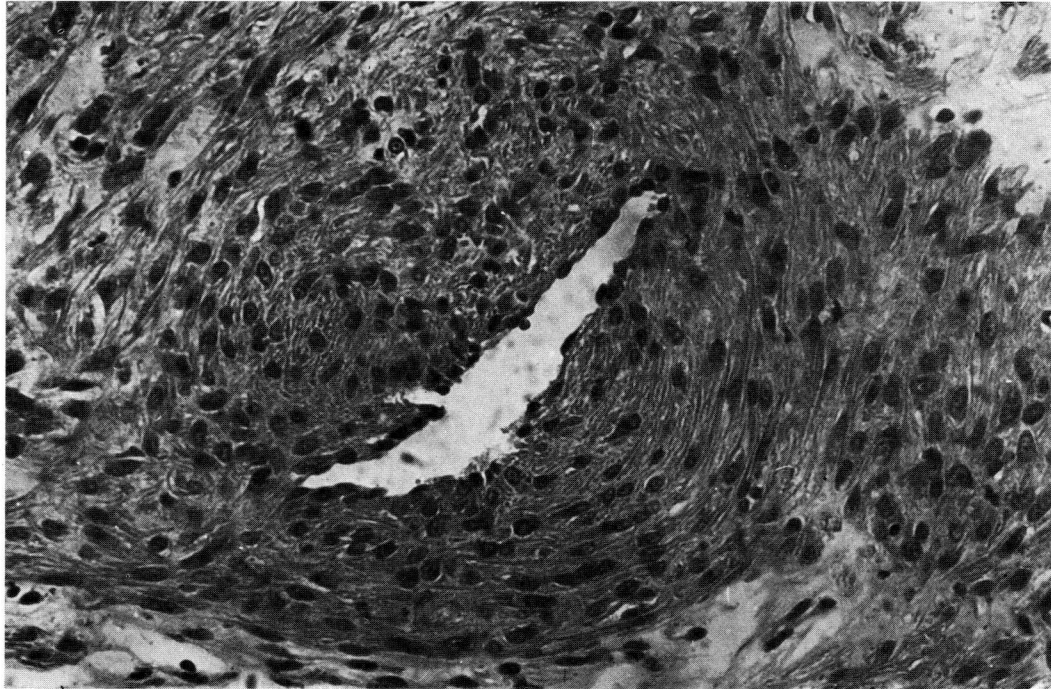

Figure 12.3
The vessels have a narrow cavity lined by an endothelium. They are surrounded by several concentric layers of 'epithelioid' cells, separated by innumerable fibrils (Pr. Bonneau) (magnified × 250).

nail, whereas the tumour may be in the pulp; hence the high failure rate of surgical treatment.

Following Iselin's recommendations, the form of treatment that we have adopted consists of a wide exposure and complete surgical excision of the lesion. The initial incision is made along the lateral border of the terminal phalanx, allowing exploration of the dorsal surface by mobilising the nail, or exploration of the pulp, where about a third of the tumours lie. The operation is carried out under local or general anaesthetic in a bloodless field. The incision must be extensive from the distal interphalangeal crease to the tip of the finger. Depending on the site of pain, the dissection is carried out up the nail bed, elevating the nail, which is replaced at the end of the procedure; or towards the pulp, until the lesion is discovered. This is generally the size of a pinhead, hard, grey or sometimes reddish in colour, and often globular.

In our experience, permanent cure is a general rule, except in those cases where long duration of the tumour has conditioned the patient to pain. Inadequate exploration is the main cause of relapse.

CONCLUSION

The extreme rarity of glomus tumours may be the reason why diagnosis is so frequently delayed; this is a most regrettable situation, since the tumour is invariably benign and amenable to surgical excision, the results of which are generally dramatic.

Familiarity with the lesion leads to early diagnosis, except in those atypical deep, multiple, or visceral forms.

BIBLIOGRAPHY

[1] AUDRY M.-C. — Nodule sous-cutané à structure de nœvus artéirel léïomyomateux. *Bull. Soc. franç. Derm. Syph.*, 1931, 222.
[2] BARRE J.-A., MASSON P. — Tumeur du glomus neuromyo-artériel des extrémités. *Bull. Soc. franç. Derm. Syph.*, 1924, 148.
[3] BEATON L.-E. — Analyse de 271 cas de T. G. *Quart. Bull Northw. Univ. med. Scho.*, 1941, 245–254.
[4] BLANCHARD A.-J. — Pathologies des T.G. *Canad. med. Ass. J.*, 1941, 357–360.
[5] BONNARD. — Sur un cas de T.G. d'un doigt de la main. *Soc. Anat. Clin. Bordeaux*, 1934.
[6] BONNET P. — T. sous-cutanée douloureuse: T. du glomus neuro-myoartériel. *Lyon chir.*, 1927, 718–721.
[7] BOZETTI F. et coll. — Report of 11 cases of PH and a review of the literature. *Tumori*, 1974, *60*, 25–32.
[8] BUNNEL S. — *Surgery of the hand.* Philadelphia, J.B. Lipinncott, 1956.
[9] CARTENSEN J. — Sur les T. unguéales. *Arch. franç. clin. Chir.*, 1927, 409–431.
[10] CHANDELUX A. — Recherches histologiques sur les tubercules sous-cutanés douloureux. *Arch. Physiol. norm. path.*, 1882, 639–682.
[11] CHEVREMONT — *Histologie.* Paris, Masson.
[12] CLARK E.-R., CLARK E.-L. — Observations on living arterio venous anastomoses as seen in transparent chambers introduced into the rabbit ear. *Amer. J. Anat.*, 1934, 229–407.
[13] CURTILLET E. — Anastomoses artério-veineuses (glomus neuro-vasculaire de Masson). *Ann. Anat. path.*, 1939, 327–345.

[14] DUNLON J. — Surgery: Primary H.P. of bone. *J. Bone Jt Surg.*, 1973, *55-B*, 854–857.

[15] FABRE L., VOISIN E. — Le glomus neuro-vasculaire et ses tumeurs. *Montpellier méd.*, 1945, 286–298.

[16] FOURESTIER J. et coll. — A rare tumor of the lip H.P. *Rev. Stomat.* (*Paris*), 1967, *68*, 553–557.

[17] GENSBER S. et coll. — Giant benign HP. functioning as an arterio-veinous shunt. *JAMA*, 1966, *198*, 203–206.

[18] GIOFFRE M. et coll. — Histopathological and clinic considerations on HP. (presentation of clinical case). *Clin. Otorinolaring.*, 1967, *19*, 335–343.

[19] GRANT R.-T., BLAND E.-F, — Observations on arterio-veinous anastomoses in human skin in the bird's foot with special reference to the reaction to cold. *Heart*, 1934, *15*, 385.

[20] IGLESIA de la TORRE, CAMARGO M., PALACIOS. — Considérations cliniques, anatomiques, radiologiques et chirurgicales de la tumeur glomique de Masson. *Cirug. orthop. Traum.*, 1939, *8*, 11–17.

[21] ISELIN M., ISELIN F. — *Traité de chirurgie de la main*. Paris, Ed. Méd. Flammarion, 1967.

[22] ISELIN M. — *Chirurgie de la main*. Paris, Masson, 1955, 572.

[23] LERICHE. — *Chirurgie de la douleur*. Paris, Masson, 1936.

[24] LORD G., DUPONT J.-Y. — Tumeurs glomiques sous-un-guéales. *Nouv. Presse méd.*, 1974, *3*, n° 8.

[25] MESCON H., HURLEY H.-J., MORETTI G. — The anatomy and histo chemistry of the arterio anastomosis in human digital skin. *I. in Vest. Dermat.*, 1956, *27*, 133.

[26] MOUCHET A., VILAIN R. — A propos des tumeurs glomiques. *Sem. Hôp. Paris*, 1949, *25*, 2771–2775.

[27] MURAD M.-T. et coll. — Ultrastructure of a H.P. an a glomus tumour. *Cancer*, 1968, *22*, 1239–1249.

[28] RIVELOS M., CODAS Q. — Considérations cliniques, anatomiques et radiologiques, et chirurgicales de la tumeur glomique de Masson. *Cirug. orthop. Traum.*, 1939, *8*, 11–17.

[29] RYPIUS E. — The roentgenologics aspects of sub-ungueal glomus tumor. *Amer. J. Roentgenol.*, 1941, *46*, n° 5.

[30] STOUT, MURRAY, BIANCHI et coll. — *Ann. Derm. Syph.* (*Paris*), 1968, *95*, 269–284.

[31] TRIFAUD A., BUREAU H. — La tumeur glomique. *Marseille chir.*, 1958, n° 1.

[32] TRIFAUD A., BUREAU H. — *Tumeurs bénignes des os et dystrophies pseudo-tumorales*. Paris, Masson, 1959.

13. THE NAIL IN RADIODERMATITIS

M. Pierre

Doctors have the doubtful honour of being the victims of practically every case of radiodermatitis of the hands. Surgeons, who use X-ray control during their operations, are most prone; radiologists are more aware of the danger of prolonged exposure to ionising radiation.

Radiodermatitis of the hands involves serious nail lesions, which are dystrophic and are due to atrophy of the matrix, which is particularly sensitive to ionising radiation. What happens to the matrix is reflected in the nail: once the matrix is injured, there is no way of preserving the normal appearance of the nail.

In the first stage of the disease the nail is covered with longitudinal striations and becomes thin and brittle. The surrounding skin is telangiectatic. As the lesions become worse, the nail starts to bulge and turn blackish. Hyperkeratosis appears under the distal edge and elevates the nail from its bed (Plate 62). Later ulceration appears around the nail bed and sometimes even on the fingertip and the pulp. When the condition is fully developed, the nail falls off (Plate 63). These nail deformities are not due only to the effect of the ionising rays, but also partly to secondary infections and mycoses, which develop in the nail tissues, when their resistance to infection is lowered.

The final stage of development is malignant degeneration, although the word 'final' is perhaps inexact, since this can in fact occur around the nail bed, when the nail itself is relatively undamaged (Plate 64).

Since nail deformity is an early feature of radiodermatitis, its appearance should warn the surgeon to stop exposing his hands to radiation immediately. In the early stage, treatment may have some effect, but once the radiodermatitis is established, surgery is necessary and consists of excision of the affected skin and a free graft, as described by Lagrot (1974). Partial or full thickness skin graft may be used according to preference; we prefer the latter. Full thickness skin flaps taken from the abdomen give a better result functionally than split skin grafts taken from the thigh or the arm. We have seen proof of this in two surgeons whose hands were operated upon within a few months of each other, using split skin grafts on one hand and full thickness grafts, taken from the abdomen, on the other. The latter gave a much better result from the functional point of view and healing was more rapid.

Careful histological examination of all the tissues removed must be done to exclude the possibility of malignant degeneration. Whether or not the nail is removed depends on the extent of its involvement, and whether or not it retains any useful function.

Replacement of the telangiectatic, atrophic, sometimes ulcerated skin on the back of the finger with a thick graft of good quality can improve the trophic changes of the nail. On the other hand, if the nail is very deformed, elevated by hyperkeratosis for example, it is better removed, being careful to remove all the matrix to prevent regrowth of irritating and painful spurs.

Even after a thorough excision of the lesion has been done, and examination of the removed tissue has shown no evidence of malignancy, this may still occur at the edges of the graft. In this case, amputation is necessary, which at least solves the problem of what to do with the nail. An over-zealous desire to conserve the nail may result in pathological tissue being left under the nail, which may be a source of future malignant change.

BIBLIOGRAPHY

[1] LAGROT F. — *Radiodermite des mains*. Paris, Doin, 1974.

14. CRUSH INJURIES OF THE DIGITAL EXTREMITIES

J. Michon and J. P. Delagoutte

This chapter deals with contusion of the fingertips, where there is sufficient retention of blood supply to allow conservation although, of course, local devitalisation of tissue will be present. In the great majority of cases these injuries have been caused by closing doors or windows, less commonly by machinery such as a press. Children are particularly susceptible to this type of accident, car doors being a common culprit.

The constant feature is bruising of the soft parts of the terminal phalanx, frequently with lacerations of the pulp or posterior structures, occasionally with both. A dorsal wound in the region of the matrix can be associated with elevation of the base of the nail from the inside outwards (Fig. 14.1). A fracture of the

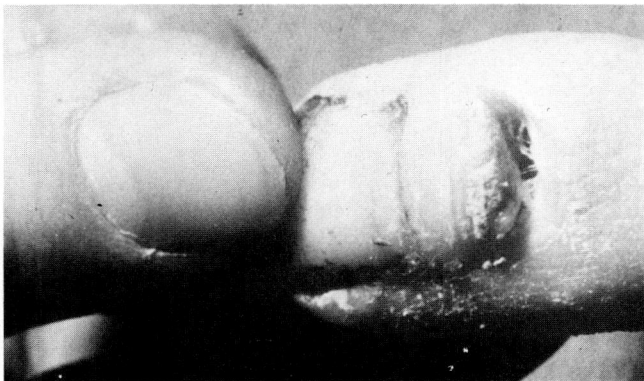

Figure 14.1
A typical wound on the back of the finger resulting from the lever action of the crushing injury on the nail.

terminal phalanx is common and may take the form of a transverse fracture of the middle part, a comminuted fracture of the distal part, or separation of the terminal tuft. Lesions of the nail are similarly variable and are responsible for the most important long-term consequences of these injuries; they are also the most difficult to treat.

Avulsion of the nail leaves the nail bed exposed and, if this is uninjured, the newly formed nail will grow normally. If the nail bed is scarred, it is likely to leave surface irregularity, which will deform the new nail, or even arrest its regeneration. These possible outcomes, the importance of which has been stressed by Iselin and Recht, can be minimised by replacing the nail or by using the nail of the big toe as a prosthesis.

Lesions of the matrix are far more serious; tearing of the anterior part will not necessarily prevent regrowth of the nail, if the base of the cul-de-sac is intact, but the scarring will produce a deformity of nail growth. Wounds extending through the matrix

will, even with careful suturing, result in at least a longitudinal groove in the nail or at worst a split nail, in which there is a tendency for the nail to loosen along the line of separation. Finally, major contusions of the matrix may lead to loss of function and failure of nail regrowth, which is a lesser problem than disordered growth, associated with scarring.

The treatment of crush injuries must take all these different elements into account. After clinical and radiological examination and surgical exploration of the wound, the devitalised tissues are excised, taking into account the good viability of pulp tissue and the desirability of conserving the maximum bone length for the maintenance of function and cosmetic appearance.

As far as the fracture is concerned, a large fragment may be controlled by splintage or by steel wire or a nylon suture; often the nail itself provides sufficient splintage to ensure adequate apposition after cutaneous suture.

This still leaves the difficulty of treating cases, in which there has been, in addition, damage to the nail matrix. Repair of a simple lesion with a fine suture, suture of the periosteal bed, and preservation of the remaining nail as a prosthesis within the matrix and on the nail bed for several weeks are the most important points. The use of the nail as a prosthesis deserves attention. It provides a good quality fixation and controls and shapes the subsequent scar formation (Fig. 14.2). If the damage

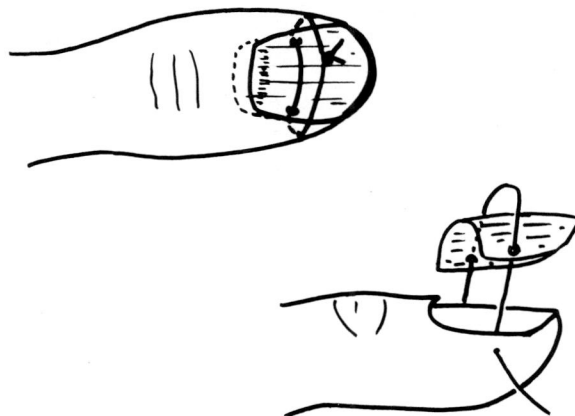

Figure 14.2
The technique for restoration of the nail that has been avulsed. Before this is done, the base and sides of the nail must be reshaped (Iselin and Recht).

to the matrix has been such that normal regrowth of the nail is beyond hope, then complete removal of the matrix by surgical excision is the wisest course. However, in children the capacity for regeneration is so great that a more conservative approach may be adopted than for adults (Fig. 14.3).

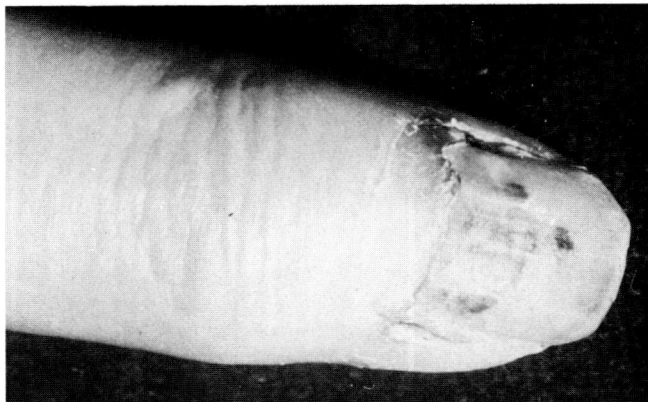

Figure 14.3
After the nail has been repositioned it is gradually pushed off by the new nail, whose bed it has been preparing.

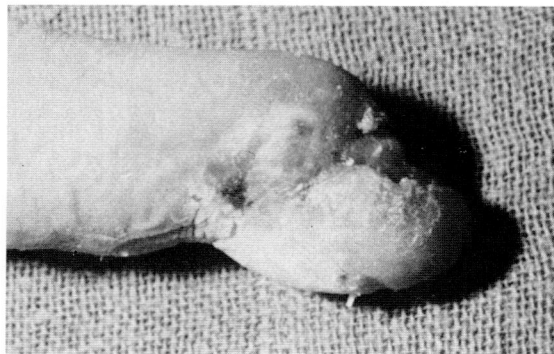

Figure 14.5
Septic arthritis following a crushing injury with open fracture of the terminal phalanx. Resection of the joint surface is necessary. The initial management of this injury failed to immobilise the lesion adequately.

Skin closure does not pose too difficult a problem in most cases. The viability of shreds of pulp tissue, even with a narrow pedicle, is remarkably good. Grafting is sometimes required on the dorsal aspect, but usually is the secondary procedure, when the skin necrosis has become demarcated in situations where it is necessary to protect tendons and joints.

Complications are necrosis and infection, which require prompt vigorous action. Skin necrosis is treated by excision and grafting (Fig. 14.4), and infected lesions by wound excision,

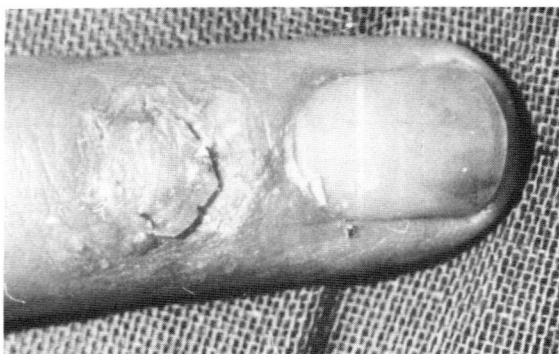

Figure 14.4
Complete skin graft following skin necrosis overlying the distal interphalangeal joint.

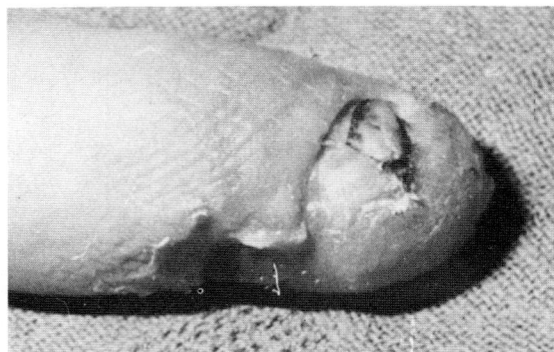

Figure 14.6
The same as in Figure 14.5, after the infection has subsided and the terminal joint has ankylosed. The deformed growth of the new nail remains an important problem.

followed by loose approximation of the skin. Delay in treatment may result in involvement of the tendon or joint, which can develop rapidly in the confined space of the fingertip (Figs. 14.5 and 14.6).

The long-term complications may involve any of the tissues and are varied: severe scarring may occur and deformity of the pulp may justify later plastic surgery; pseudarthroses is rare, but exostoses may form, which may require removal if they deform the nail.

The long-term consequences of nail damage are the most frequent and difficult to treat. In some cases, the most reasonable solution is excision of the matrix and our experience

does not agree with Iselin's pessimistic view. We have not found a deficiency of pulp sensitivity after the period of tenderness that follows any intervention has subsided. We use this radical method widely in manual workers, but it must be stressed that this is not a simple procedure and must be done with precision. For deformed nails in which the matrix is still healthy, avulsion of the nail and correction of the bed by positioning a toe-nail in its place is a method which can give good results, if the prosthetic nail is left for a sufficient length of time to assure restoration of the nail bed and to allow regrowth of the new nail.

Persistent pain, without obvious anatomical explanation is outwith the scope of surgical treatment. Rehabilitation towards desensitisation overcomes the problem in most, but the psychopathological element of pain must always be remembered and recourse taken to specialist help, if necessary.

As in all traumatology, in the planning of the management of crush injuries to the fingertip it is necessary to take account of the long-term consequences. The prevention of complications and poor cosmetic appearance, so badly tolerated in the fingers, demands good repair and good judgement with regard to sacrifice and conservation of damaged tissues.

15. AVULSION INJURIES OF THE NAIL

M. Iselin

This subject was reviewed by P. Recht and J. Bazin (1961) and in our *Traité de Chirurgie de la Main* in 1967 with François Iselin. I am not therefore going to reiterate the findings, but point out what still holds true in 1980 and what has not endured the test of time.

The existence of sensory problems, amounting even to anaesthesia of the pulp after the nail has been destroyed, has been confirmed. This, verified by Moberg's test on old injuries, has formed the basis of our work, which has led to the concept of the importance of the nail in crush injuries of the fingertip.

Bazin has also verified the existence of this surprising anaesthesia in recent injuries, as well as its progressive disappearance following nail regeneration. However, we have come to realise that this is not a constant phenomenon. Some fingers, where the nail has been lost, are anaesthetic, some merely hypo-aesthetic, and others show no long-term changes in sensitivity.

TREATMENT OF AVULSION OF THE NAIL BY REPLACEMENT OF THE NAIL

In a crushed finger, avulsion of the nail is always accompanied by damage to the nail bed and sometimes by fractures of the terminal phalanx. Studying the pathology of this lesion, our attention was drawn to the laceration of the nail bed. When this was transverse and distal, the new nail grew up to the level of the laceration but beyond this the nail bed was covered by skin. When the lacerations were multiple, the regrowing nail was either split in two or arrested at islands of scar, which had to be removed in a secondary operation.

Regrowth of the nail can be modified by two factors: damage to the matrix, which is the source of new growth and lesions of the nail bed, which prevent correct alignment and guidance of growth. It is imperative that the scarring of the nail bed should be such that its function as a guide is not impaired. Unfortunately, the bed is too thin and friable to be sutured. We therefore thought that by covering it with a nail, which protects and maintains its shape, distortion by scarring might be prevented. The use of the nail in this way is of major importance and other methods of splintage are only of secondary importance. These principles remain important now, but the long-term results are not always favourable with regard to the completeness of growth. When the replaced nail remains adherent to the bed, regrowth is perfect and the site of injury unrecognisable in under a year. However, when the nail is shed, regrowth is less satisfactory and often ceases at the level of the original injury.

In our first series, we were surprised to see the replaced nail adhere like a graft. This is, unfortunately, not the rule and more often the nail tends to separate from the bed, falling off spontaneously around the twelfth to fifteenth day.

Despite the inconstant results, we consider that the replacement of the nail is indispensable. The nail affords protection from pain, acts as a splint and obviates the need for external splintage, which provides poor immobilisation.

Our operation is carried out as a delayed procedure so that tissue viability may be assessed, and in order to avoid amputations which might, at first sight, have appeared necessary. Two to four days after the injury flaps of tissue, which may have had a very narrow pedicle, may appear entirely viable. In addition, if one wishes to replace the nail, there is great danger in covering an open fracture at the same time. Clinical experience has shown that this is not a purely theoretical danger.

The operation is done in two stages: an initial cutaneous repair with minute sutures, paying particular attention to restoring the shape of the nail bed; and the delayed replacement of the nail. In the best cases the nail adheres and ensures growth of a normal nail. It must be trimmed to fit the nail bed, otherwise it might ulcerate the nail folds. If the nail has been completely removed, it is stored for the two to four days pre-operatively in mild antiseptic solution, after which it is tailored and put back in position. When the nail has been lost, we either use a nail from our bank, or, if this is not available, a skin graft. The nail is held in place by two transverse stitches which fix it perfectly. No post-operative immobilisation is required, minimising stiffness and delay – no splint, less plaster, a simple dressing. Soft tissue healing is complete at ten days. Complete healing takes an average of thirty days, but some patients resume work at twenty-five days.

CRUSH INJURY WITHOUT FRACTURE

This injury is particularly common in children for whom removal of the terminal phalanx is the therapy all too often prescribed. The initial appearance may be discouraging: ragged flaps of skin hang off, revealing the denuded terminal phalanx which points through the pulp. The nail may be detached or still adherent. However, after two to four days, the picture is quite different, and the most unpromising injuries may appear viable, allowing the second stage of the procedure to be carried out (Fi 15.1).

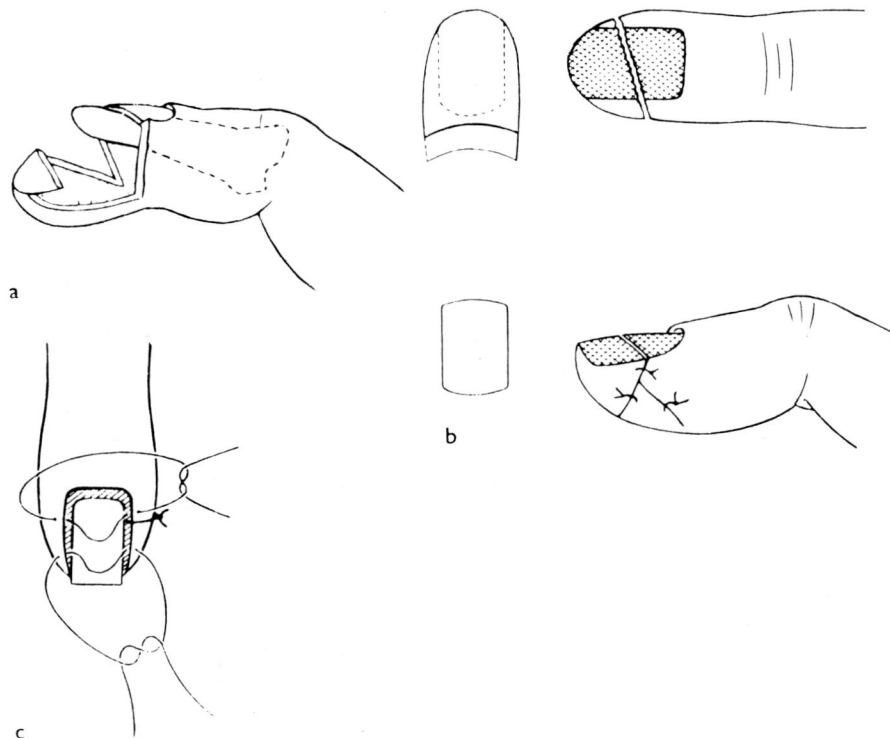

Figure 15.1
Diagram showing the lesion and how it is repaired.
a   The nail, its bed and the sides of the pulp are lacerated, only a narrow pedicle is left to ensure vascularisation;
b   The first stage of treatment: the skin is sutured and the nail reshaped as indicated by the dotted line;
c   Stage two: the nail is put back on the nail bed; it is kept in place by two stitches (under which a small pledget may be put). When the nail
    is put back in this way, it helps to ensure correct immobilisation and regrowth of a healthy, new nail.

(Extract from *Traité de Chirurgie de la Main*, Flammarion Médecine Sciences, 1967.)

CRUSH INJURY WITH FRACTURE OF TERMINAL PHALANX

A similar procedure is followed. The fracture will be splinted by the nail and may otherwise be ignored, apart from removal of detached fragments from the wound in the nail bed. Internal fixation by a pin is required only when there is a displaced distal fragment.

SKIN GRAFTING OF RAW SURFACES

This is mentioned as it is the current American technique. We find it to be of use in two situations: when there is no nail to replace and when there is extensive skin loss on the dorsum of the finger which requires grafting. If the nail regrows, it separates the graft from below and its appearance is as normal as the scarring of the nail bed will allow.

It is important to adhere to the principles outlined above. Immediate covering of an open fracture of the terminal phalanx with nail or graft courts too many dangers.

16. NAIL PROSTHESES

C. Dufourmentel

Destruction of the nail is a conspicuous and unaesthetic mutilation, often difficult to accept psychologically, particularly in young women and girls. We commonly perform a purely cosmetic operation for an extremely dystrophic nail, which has regrown as a hypertrophic horn or multiple spurs, or for a mutilating injury to the finger with amputation and loss of germinal matrix.

McCash, who has studied the problems extensively (McCash, 1964) describes two reparative procedures: the first consists of a free graft of a toe-nail with its nail bed, a thin bony layer and its matrix, as a unit to the finger. R. Mouly has used this method with some success (Mouly and Debeyre, 1961). The second method consists of the creation of a bed and setting for application of a prosthesis giving as normal an appearance as possible. False nails are available commercially and used by women to cover nails, which may be short or damaged by housework. These prostheses are aesthetically satisfactory, particularly if covered with nail varnish. Some are self-adhesive, others are sold with a special non-irritant glue which fixes them to the natural nail.

These prostheses cannot be used directly on a mutilated nail because of the irregularity of the bed, where the nail is dystrophic and fragmented and there is loss of nail folds. The thin fold of epidermis, which covers the base of the nail, and the proximal parts of the lateral folds are very characteristic in the normal nail.

In six cases we have attempted reconstruction of both the nail folds and nail bed at the same time in order to fit a prosthesis. We followed, with slight modification, McCash's technique. Three of our cases gave satisfactory, if not perfect, results, which encouraged us to publish details of the techniques.

PREPARATION OF THE NAIL BED AND REMOVAL OF THE MATRIX

First, the desired contour of the nail is outlined and, within that area, all the residual skin and scar are removed. A little soft tissue is left on the bony surface, if possible, to provide a satisfactory surface for nutrition of a graft. At the base of the nail the remaining matrix is resected from beneath the skin, and a subcutaneous flap mobilised for at least one centimetre proximal to the apparent nail base. The skin is detached further in a lateral direction. It is this flap which forms the seating for the prosthesis.

CREATION OF NAIL FOLD

A split skin graft is taken from the thigh or buttock to avoid visible scarring. This graft is applied to the nail bed and pushed beneath the subcutaneous fold, to cover both the deep and superficial surfaces of the plane of separation. To maintain the graft in position, a prosthesis is inserted immediately. It is simple and less traumatic to use the greaseproof paper which separates layers of grease-impregnated tulle. This paper can be trimmed to the desired shape, covered with the graft (its raw surface facing outwards) and buried in the fold proximally and laterally; a layer of tulle and compressive dressing keep it all in place and it is retained for five days.

It is vital, then, to avoid the collapse of the fold. To this end, a new piece of greaseproof paper, which has been trimmed exactly, must be inserted for a further period. Around the fifteenth day, it can be replaced by the definitive prosthesis for part of the day and put back in at night. After six weeks, the prosthesis may be worn alone. Unfortunately, it requires to be stuck down, which caused two of our failures, because the patients could not tolerate the glue. We thought surgical varnish the best product.

In our six cases, we observed two secondary failures due to intolerance of the glue with progressive effacement of the nail setting. One case was a primary failure with infection leading to rejection of the graft, this patient refusing a second operation. The other three gave satisfactory, if not perfect, results (Figs 16.1, 16.2 and 16.3).

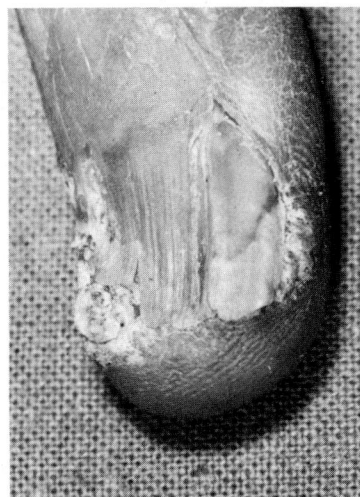

A

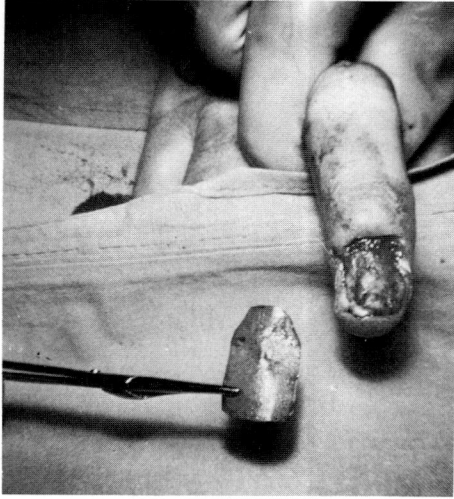

B

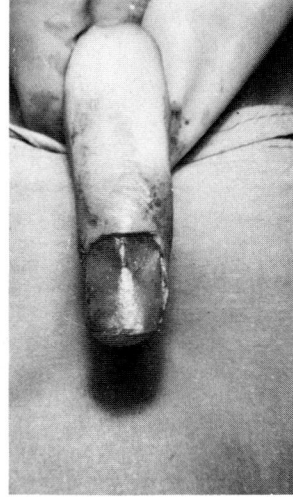

C

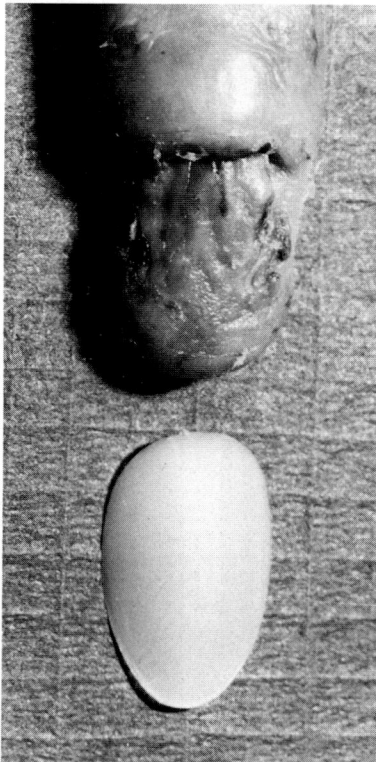

D

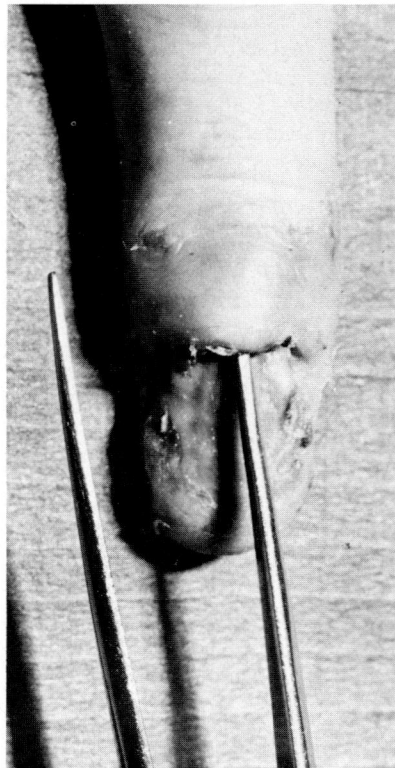

E

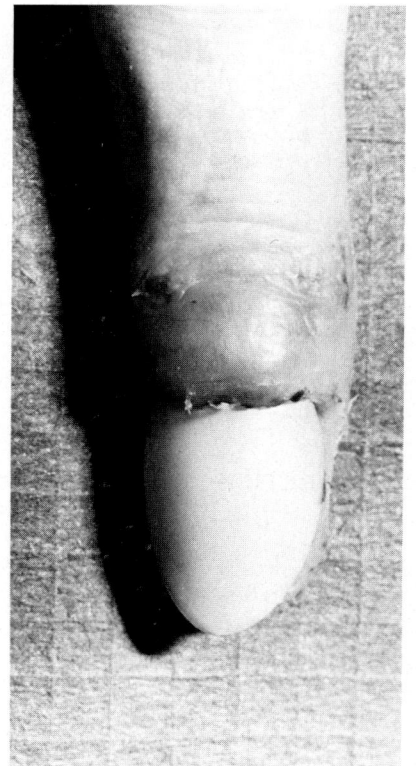

F

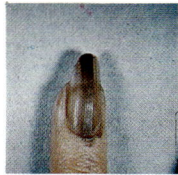
53

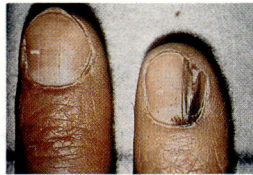
54

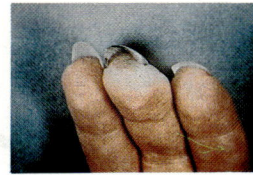
55

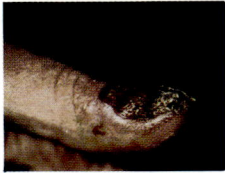
56

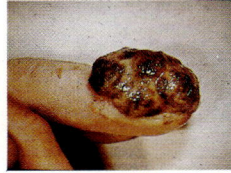
57

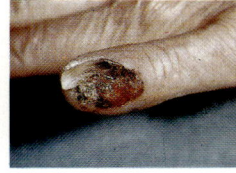
58

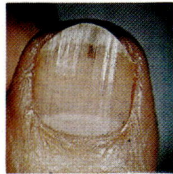
59

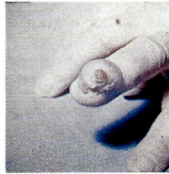
60

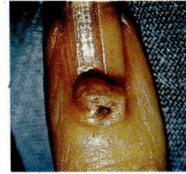
61

Plate 53 Central subungual malignant melanoma

Plate 54 Lateral subungual malignant melanoma

Plate 55 Melanotic whitlow involving nail and pulp

Plate 56 Melanotic whitlow forming a tumour with bursting of the nail

Plate 57 Amelanotic subungual malignant melanoma with loss of the nail

Plate 58 Amelanotic malignant whitlow

Plate 59 Benign subungual melanoma

Plate 60 Subungual kerastosis

Plate 61 Subungual exostosis

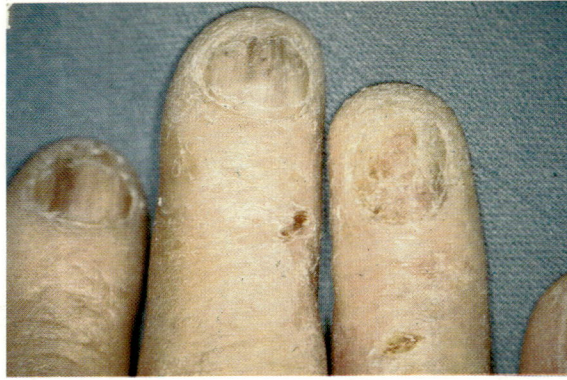

62

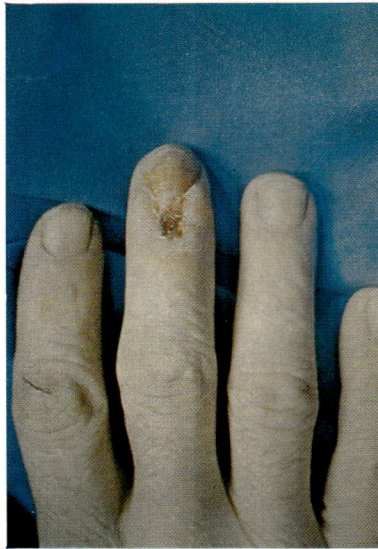

63

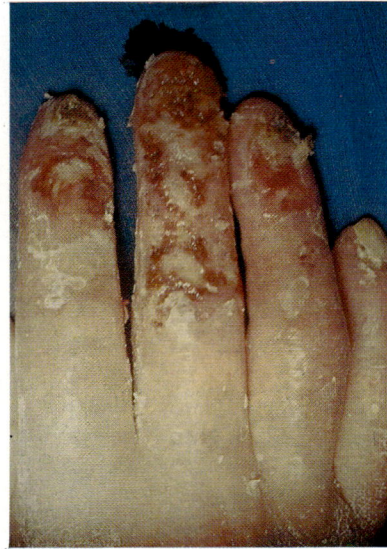

64

Plate 62 Deformation of the nails, which are curved and striated with blackish coloration. Hyperkeratosis is present under the tip of the nail

Plate 63 Radiodermatitis with ulceration and loss of the nail

Plate 64 Squamous epithelioma has developed at the level of the nail fold, without gross deformity of the nail itself.

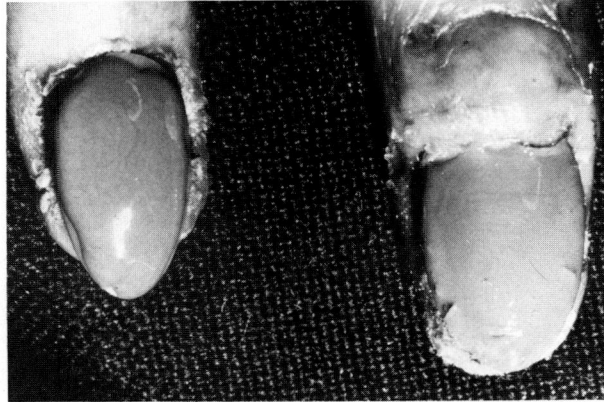

G

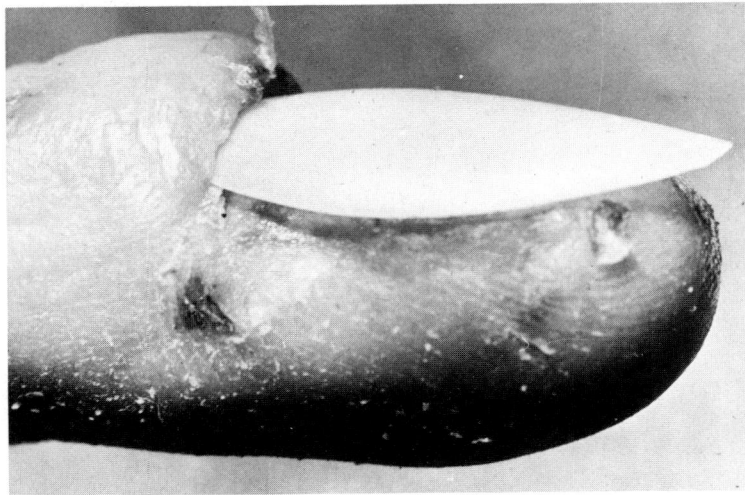

H

Figure 16.1
First case
A Left-hand middle finger-nail destroyed by infection six months previously. Exuberant crusting remains, which is ugly and also awkward because it snags readily and bleeds.
B Refashioning of nail bed and seating have been done. Piece of thick grease paper has been shaped as required and is covered with split skin graft, with raw surface to outside.
C Graft and its support are placed on finger and introduced into the proximal seating. Unfortunately, the state of the lateral skin is too bad to allow formation of deep lateral folds.
D Nail bed and prosthesis six weeks later.
E Depth of pouch at base of nail is shown.
F Prosthesis in place.
G Adjacent finger beside prosthesis.
H Prosthesis in place from side. Note that lateral seating is inadequate.

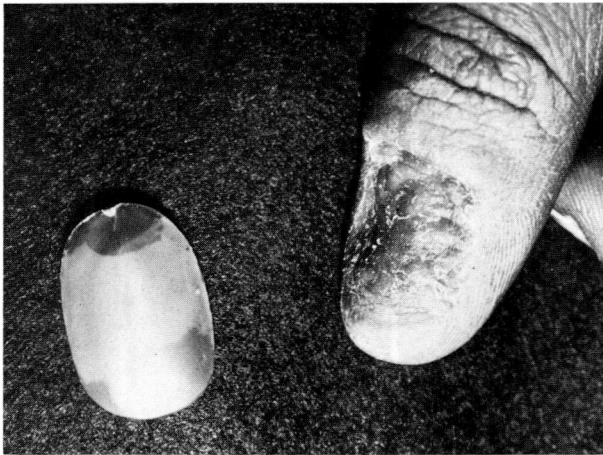

A

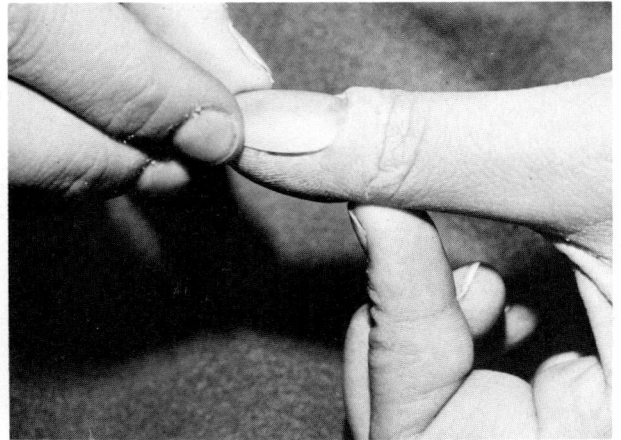

C

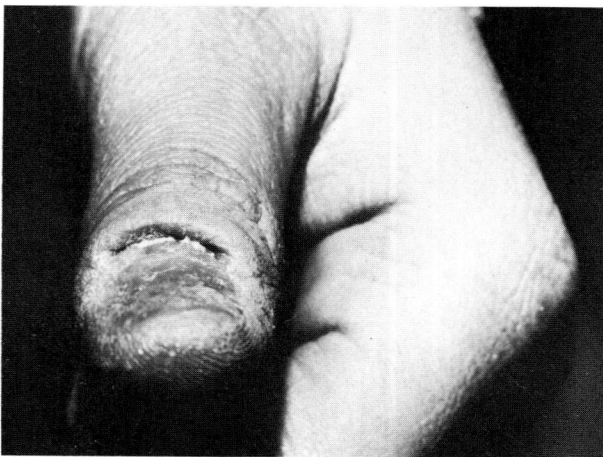

B

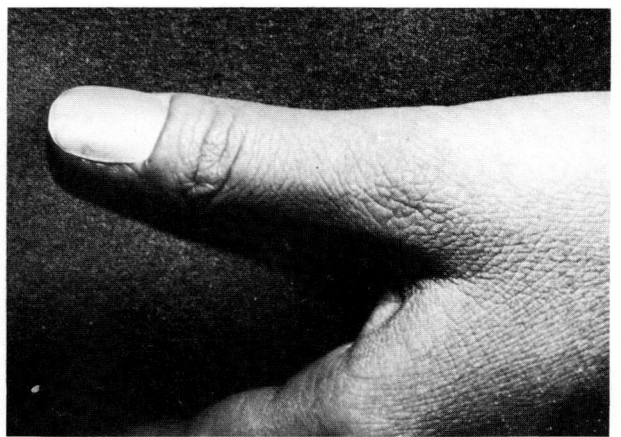

D

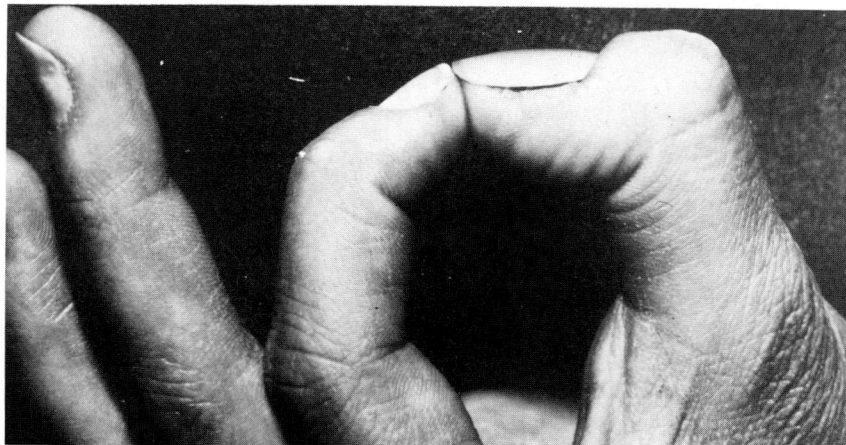

E

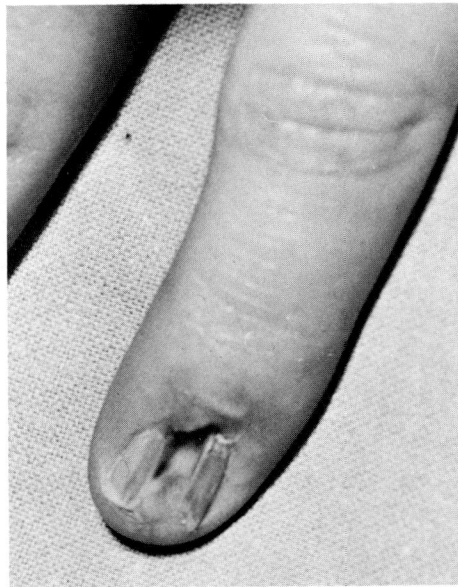

A

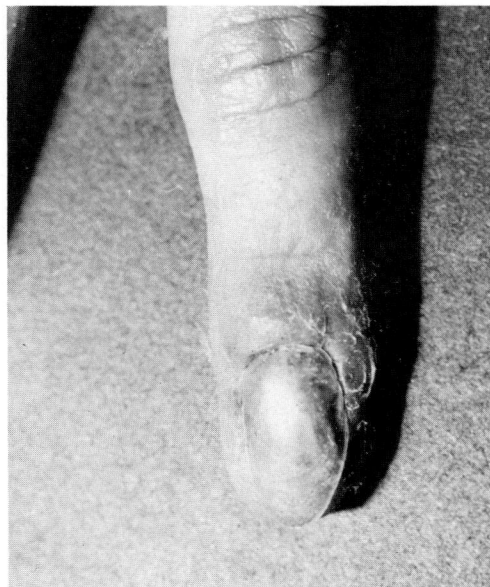

C

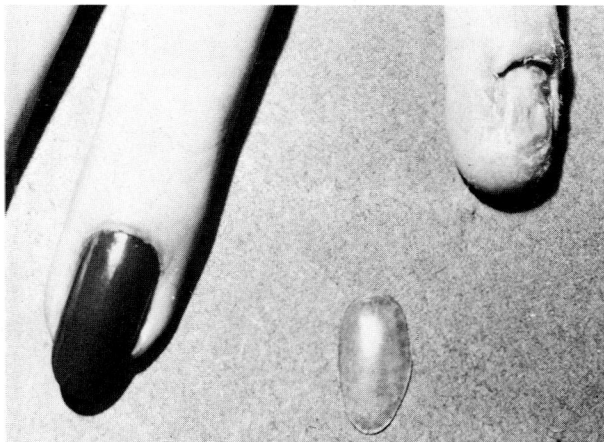

B

BIBLIOGRAPHY

[1] BUNCKE H. J. J., GONZALES R. J. — Finger nail reconstruction. *Plast. reconstr. Surg.*, 1962, *30*, 452.
[2] McCASH C.-R. — Treatment of finger nail deformities. *Proceedings of the third international congress of Plastic Surgery. Washington (D.C.) October 1963*. Amsterdam, Excerpta Medica, 1964.
[3] MOULY R., DEBEYRE J. — Le gigantisme digital – étiologie et traitement. A propos d'un cas. *Ann. Chir. plast.*, 1961, *6*, 187–194.

Figure 16.3
Third case
A Nail bed before operation.
B Nail bed after operation.
C Prosthesis in place. In this case, lateral seating is more satisfactory but skin round operated area is a little dystrophic.

◄ Figure 16.2
Second case
A Reconstructed nail bed and prosthesis (left thumb).
B Proximal pouch seen from end of finger. Here again lateral seating is inadequate.
C Prosthesis being inserted.
D Prosthesis in place.
E Thumb seen from side.

17. PROSTHESES IN AMPUTATIONS OF THE FINGERTIPS

J. Pillet

In his work on the physiology of joints, Kapandji writes: 'The hand is a marvellous, versatile tool, thanks to its essential prehensile function. Opposition of the thumb to the other fingers, which is a prerogative of man, confers the ability to grip a wide range of objects.'

The smallest lesions of the hand, such as the loss of a fingertip or even a nail, will impair the function of this precision tool, and may cause psychological distress. A cosmetic prosthesis applied, once the wound has healed, will allow the patient to accept his amputation; finding that his hand looks normal, he will not be ashamed to show it and will adapt to it more readily. The preparation of a prosthetic nail is simple; its retention, however, poses several problems which are not yet resolved.

SIMPLE NAIL PROSTHESIS

Fixation is achieved with glue, or by surgery.

ACRYLIC PASTE

Commercial acrylic paste can be used if part of the nail has been preserved. It is applied with a brush or spatula. This technique is easy for an experienced manicurist, but difficult for the patient, as it has to be repeated every two or three days.

COMMERCIAL 'NATURAL' NAIL

This is available commercially to conceal broken nails. Therefore it has the disadvantage of being merely placed on the finger and not fixed in. Fixation is finicky and the end result appears bulky.

COLLARED NAIL

Here a very thin supple collar surrounds the nail, securing it firmly. This method seems simple and effective, but it does have drawbacks: the prosthesis has a variable thickness and texture that is hard to manufacture and fixation is difficult for the patient. If the fixation is secure, daily removal causes irritation, as does the maceration which develops if the prosthesis is left in place for a week. Finally, if the collar does not work well the prosthesis will not stay secure.

SURGICAL NAIL

A nail, which may be either standard or specially made for the patient, is placed in a pouch, which is lined with split skin graft.

This technique is excellent in theory, but shortly after its formation the pouch tends to close and the nail is extruded.

NAIL PROSTHESIS WITH FINGER ATTACHMENT, OR 'THIMBLE' PROSTHESIS

This is indicated in a wide variety of cases ranging from a deformed nail to complete loss of a terminal phalanx. A simple finger cot is used to cover the terminal phalanx; it sits easily on the stump but occasionally the stump may require minor surgical alteration if it is too bulky or deformed. The material of the cot must be fine and supple to maintain pulp sensitivity and must have the same marking and colourings as the finger.

The soft nail cannot be varnished and, if this is desired, it is better to replace it with a hard acrylic nail, which is fixed by its edges to the inner surface of the finger cot.

The proximal edge of the prothesis, which is very fine, extends either to the proximal interphalangeal joint, where it will be hidden by the skin creases, or to the base of the finger, where it can be hidden by a ring. Each is satisfactory and the choice depends on patient preference.

RESULTS

1. SIMPLE PROSTHESIS

We have never obtained satisfactory lasting results using these prostheses. It is difficult to satisfy a patient, who attaches great importance to the slightest detail and these techniques can never reach perfection.

2. THIMBLE PROSTHESIS

Patients are immediately satisfied with the aesthetic result, forget their amputation and will use their hand freely. Some patients report that their prosthesis, as well as being aesthetically satisfactory, plays an important functional role. The prosthesis gives them a longer finger and affords some protection, while conserving the sensitivity of the pulp.

MATERIAL USED

Since 1953 prostheses have been made from polyvinyl chloride. The disadvantages are that it turns yellow and gets dirty very quickly; it hardens and shrinks, and replacement is required every six months.

An architect had an accident with a car fan-belt in February 1965, causing terminal segment amputation of the third finger. A prosthesis was supplied in September 1966, and worn all day since. He 'put up' with his polyvinyl chloride prosthesis, but he cannot do without his silicone prosthesis, which was supplied in June 1969.

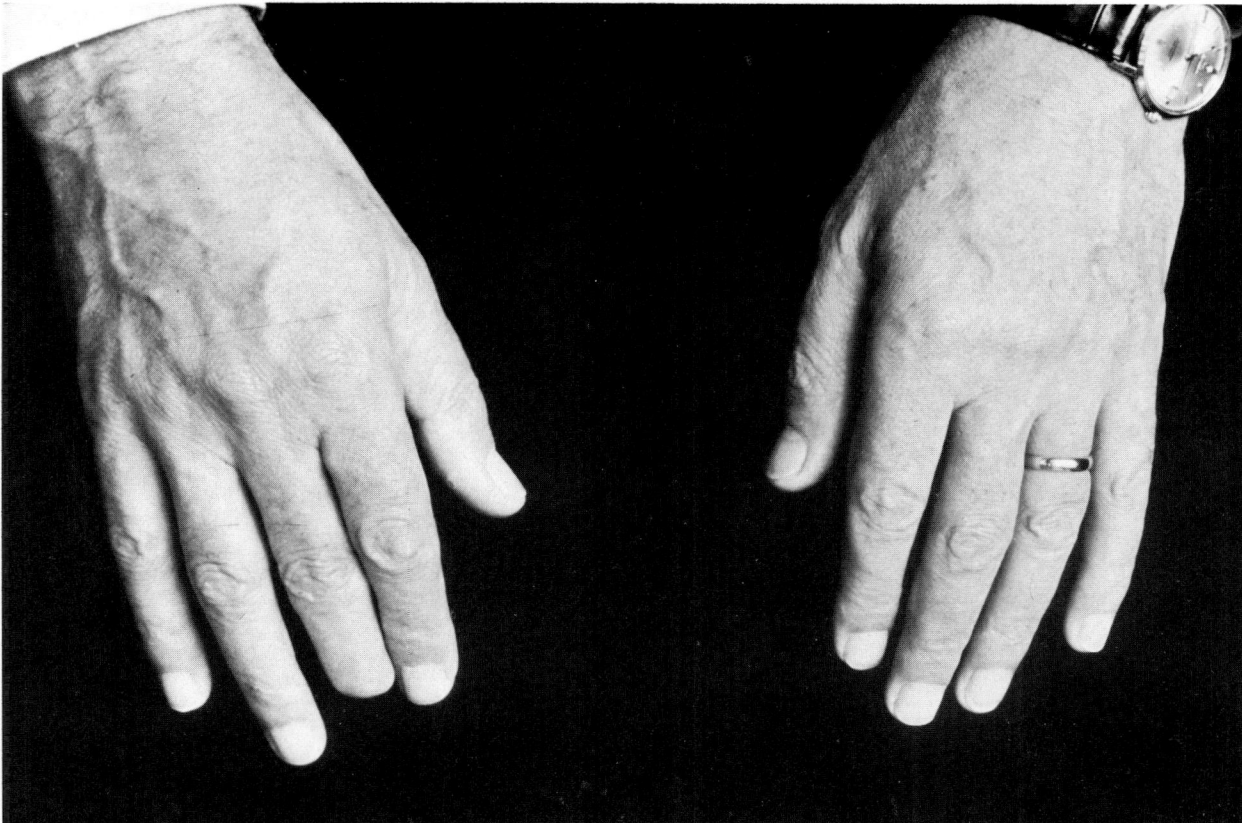

Figure 17.1
Before fitting the prosthesis.

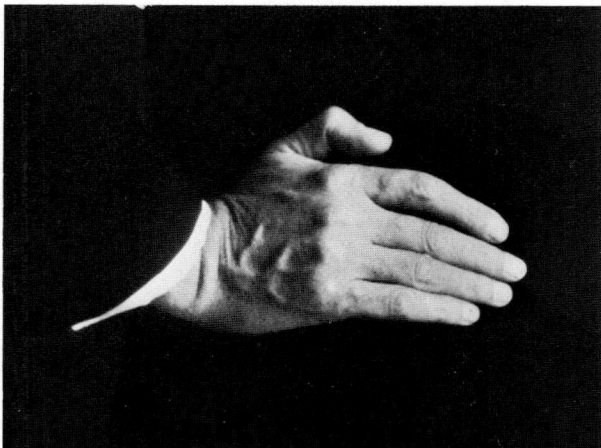

Figure 17.2
The prosthesis in place.

During the last eight years we have been using silicone with excellent results: the suppleness and colour do not change and are well tolerated by the skin.

CONCLUSION

According to the statistics supplied by the prosthetic department at the Saint-Antoine Hospital (under Professor Gosset), a large percentage of patients wear their prosthesis regularly for several months, then episodically for the next three years before finally giving it up altogether. There are others, however, who would not be able to do without them.

Whether a prosthesis is worn temporarily or permanently, it allows a patient to get used to his amputation and rejection after a period does not therefore imply failure.

Deformed nail and pulp following infection; patient wears prosthesis all day with no problems.

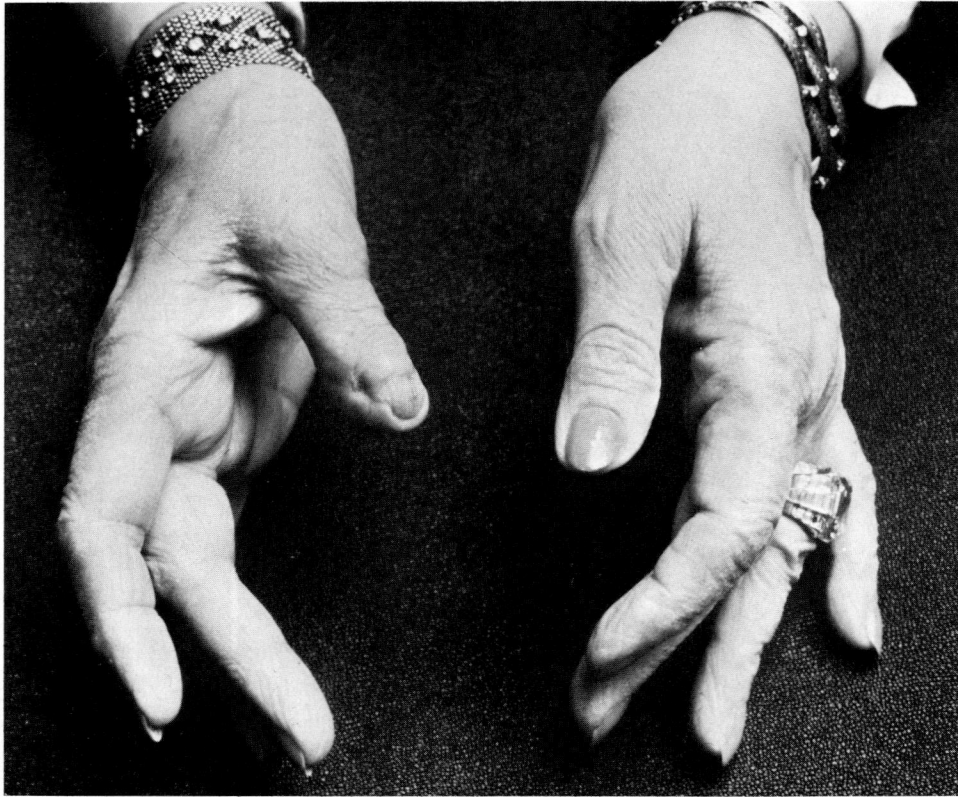

Figure 17.3
Before application of prosthesis.

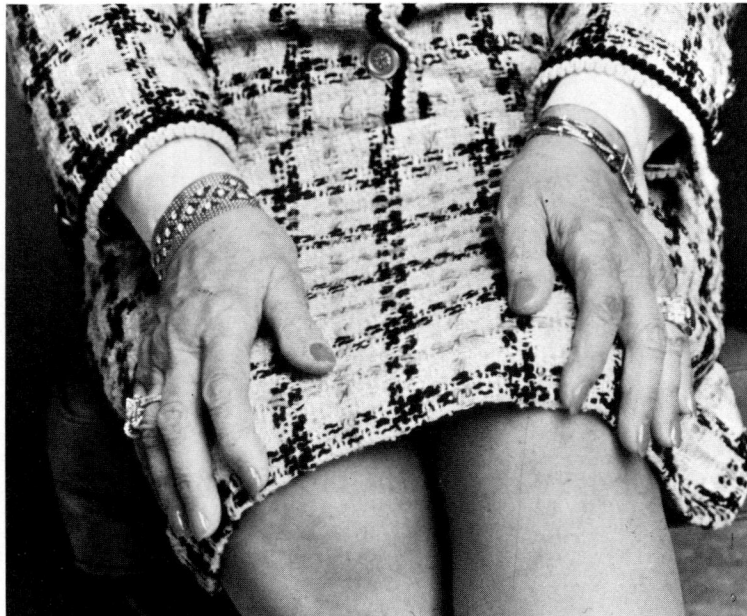

Figure 17.4
Prosthesis in place.

18. PLASTIC SURGERY AND THE CLAW NAIL

C. Verdan

Among the many cases of nail pathology that need surgical correction one of the most difficult is that of claw nail. In some amputations of part of the fingertip, all or part of the nail bed may be removed also, depriving the nail of its physiological support. When the nail grows again, instead of being parallel to the finger it gradually curves obliquely or perpendicularly to lie over the end of the finger. Sometimes the whole of the terminal phalanx may be removed, while part of the nail matrix is left. In this case the deformity is particularly pronounced.

The claw nail is not only unattractive, but it interferes with the proper function of the finger. By extending over the end of what remains of the finger pulp, it either slides over the objects the finger is trying to get hold of, or else its free edge becomes fissured and tends to snag materials. It may also be painful. As a result the finger is not used and becomes a liability. It spoils the harmony of the hand and interferes with the hand function as a whole.

There are four possible solutions:
(a) a complete excision of the nail matrix, which will prevent it from regrowing; we know, however, that aberrant spurs originating in the corners of the matrix recur frequently
(b) advancement of the finger pulp by V–Y plasty, after removal of the nail
(c) osteo-plastic elongation, which allows the nail to grow parallel to the finger
(d) recession of the nail and its matrix on a dorsal 'caterpillar' pedicle skin graft combined with a V–Y advancement of the pulp.

EXCISION OF THE NAIL MATRIX

In the doctoral thesis entitled 'Chirurgie réparatrice des pertes de substance et des amputations du pouce', written by Simonetta (1975) under the supervision of the author of this chapter, there is the following paragraph.

> 'Amputations through the nail and its bed are a special problem, because regrowth of the nail is only possible if at least a *third* of the nail bed remains intact. If this is not the case, a resection of the rest of the nail bed and *all* the matrix must be done in order to avoid an unattractive, irritating regrowth.'

This radical solution may also be required for the claw nail. However, it does cause a considerable loss of skin, which will require to be covered by a graft, and is also rather unsightly. It is tempting to cover the stump with thick skin, derived from the palmar aspect of the finger, but, if the surgeon wishes to resect the whole of the nail matrix to avoid regrowth of nail spurs, he will have to shorten the bone or even excise the entire terminal phalanx. This sacrifice of the insertions of the deep flexor and extensor tendons is regrettable, since not only will the power of flexion diminish, but the proximal interphalangeal joints will not be able to move as much as the remaining fingers. This is particularly true in the case of the fourth and fifth fingers, whose superficial flexor tendons are sometimes atrophied.

This type of operation may, however, be indicated in manual workers, who are anxious to return to work as soon as possible and are more concerned with early return of function than with appearance.

This operation of finger shortening leaves a well-vascularised stump with normal sensitivity (Fig. 18.1), which is very important for those who work out of doors, particularly in cold climates, and for those whose jobs require precise movements and the use of delicate instruments, which demand good sensitivity of the fingers. In many instances, however, a more sophisticated solution to the problem is required.

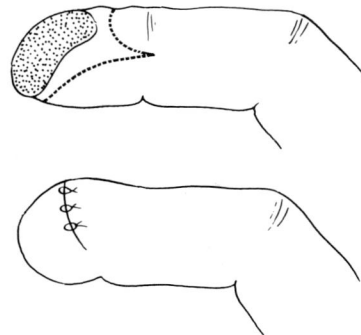

Figure 18.1
Complete excision of nail and matrix.

PULP ADVANCEMENT

In an emergency, guillotine types of amputation of the tip may be covered by bilateral, triangular flaps of skin advanced by the technique of Kutler or with a single median flap as a V–Y advancement, described by Tranquilli–Leale. These solutions avoid having to shorten the injured finger and at the same time provide the new nail with support of the soft parts, which prevents a claw nail from forming, or at least minimises the chance.

For cases of established claw nail which are not too severe, a similar solution may be carried out using secondary surgery. In these cases, part of the skeletal support is still present and only half or two-thirds of the nail bed has been removed. The nail needs only a little support to grow in the right direction and to be cosmetically and functionally satisfactory.

In these circumstances it is enough to simply remove the nail and then create a large Tranquilli–Leale V–Y flap which can be advanced as required (Fig. 18.2) to fill up the end of the pulp.

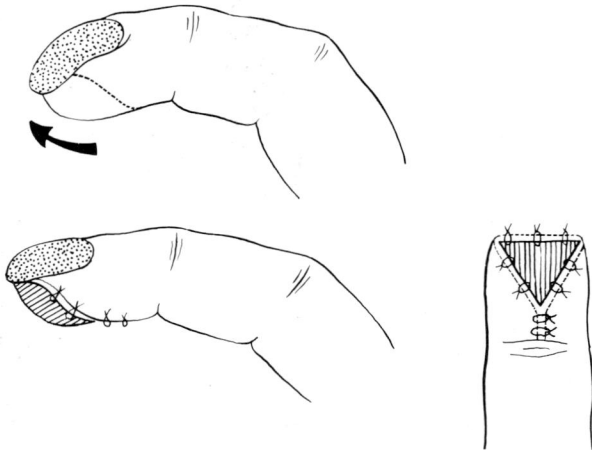

Figure 18.2
After the nail has been removed, the flap of skin from the finger pulp is brought forward in a V–Y plasty. The new nail will grow on top of it.

The nail will then grow again in a normal alignment, since it is adequately supported. The correct growth of the new nail may be helped by applying an artificial nail to the bed for a short time after the operation. This is a simple procedure which should be more widely used.

OSTEO-PLASTIC LENGTHENING

The third solution is essentially one of reconstruction. It allows the interphalangeal joint to work again normally. The procedure consists of lengthening the terminal phalanx with a shaped bone graft, covered on the palmar aspect with a cross-finger flap.

OPERATIVE TECHNIQUE

A semi-circular incision is made in the finger pulp (Fig. 18.3b), perpendicular to the axis of the finger and therefore parallel to and about 2 mm from the lateral and terminal borders of the nail. The plane of section extends to the nail bed and divides two-thirds of the palmar thickness of the terminal segment. The dorsal flap, which corresponds to the nail bed with its matrix, is preserved (Fig. 18.3c). The rich blood supply ensures that the quadrangular flap remains viable. It is gently elevated by a rugine, passed under the nail matrix, to allow full correction of the nail alignment. This correction may amount to 90° in severe deformities. The nail remains attached to its bed through its

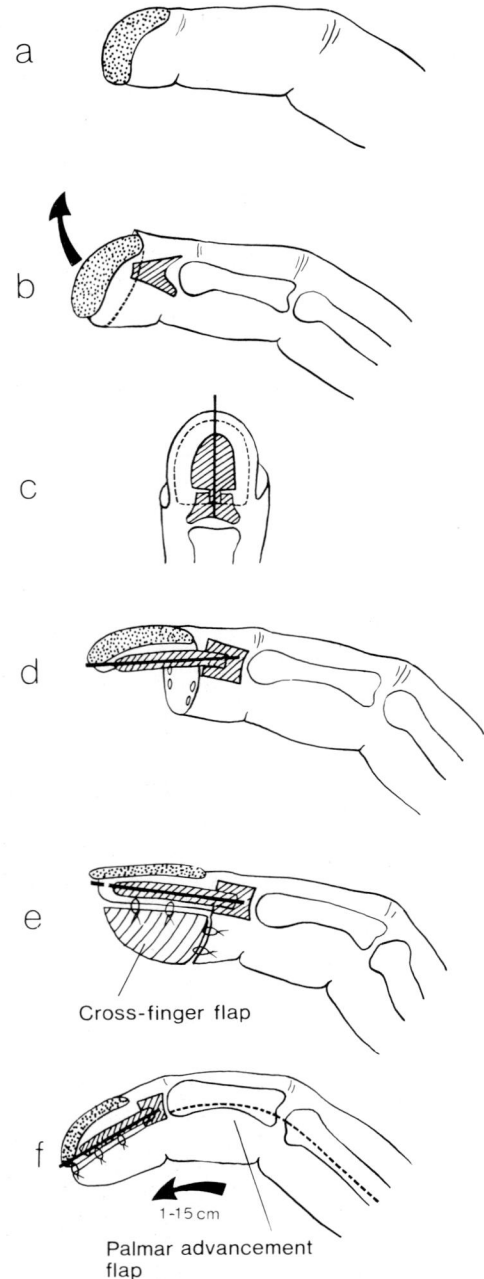

Cross-finger flap

1–15 cm

Palmar advancement flap

Figure 18.3
Osteo-plastic lengthening of the terminal phalanx with the bone graft (C, D) and a cross-finger flap (E), or a palmar advancement flap (F).

normal keratin attachments. The hard nail plate, which is slightly curved in both directions, remains surrounded by a narrow strip of skin, which will act as nail folds (Figs. 18.3b and 18.3c). In their normal state, the nail folds border the nail grooves, but in the case of claw nail the lateral grooves frequently disappear because of the deformity. When the dorsal flap has been elevated, what is left of the skeleton of the terminal phalanx is exposed and cleared of soft tissue – care being taken to preserve the two tendon insertions. The end is curetted or freshened with bone cutters. Using an air drill, a slot is then made about 2–3 mm wide, varying in depth according to circumstance. In the case of a child, care must be taken to preserve the growth cartilage. If the terminal phalanx is completely absent, the slot must be made in the head of the middle phalanx (Fig. 18.5).

Using an osteotome, a small cortico-cancellous graft is removed from the iliac crest. At this stage it is rectangular and too large, but it is shaped by means of bone nibblers to form a small paddle (Fig. 18.3c), with the handle shaped to fit the slot in the phalanx. This is then secured with fine Kirschner wire, 0·7 mm in diameter. An X-ray check may be useful at this stage. If the proximal bone fragment is very short, the wire may have to extend across the terminal interphalangeal joint in order to ensure stability of the delicate construction.

The dorsal flap containing the nail now rests directly on the cortical side of the bone graft (Fig. 18.3d), beyond which it

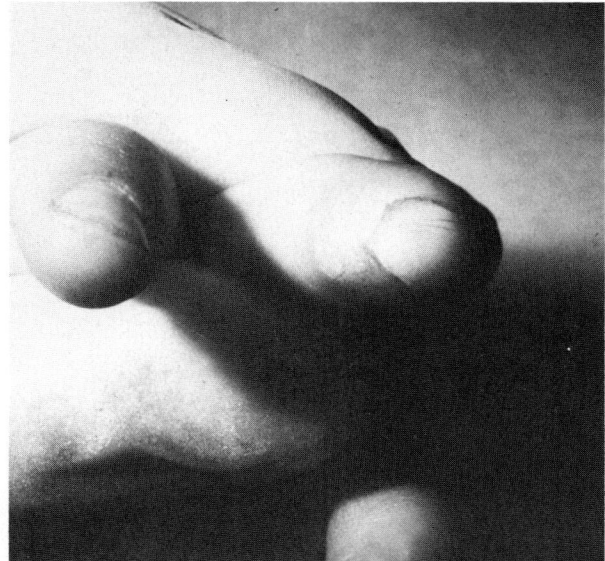

A

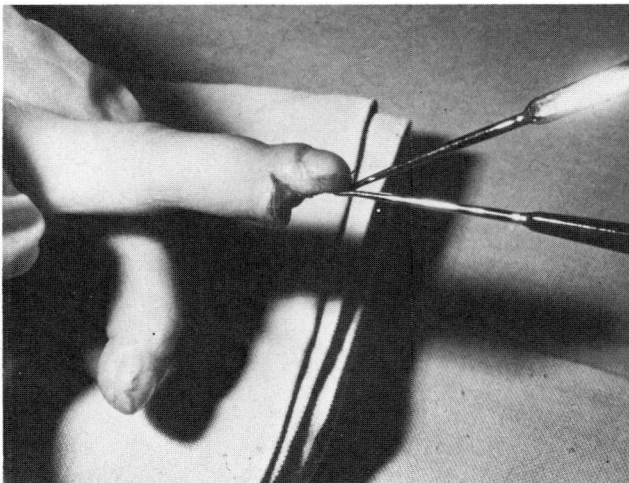

B

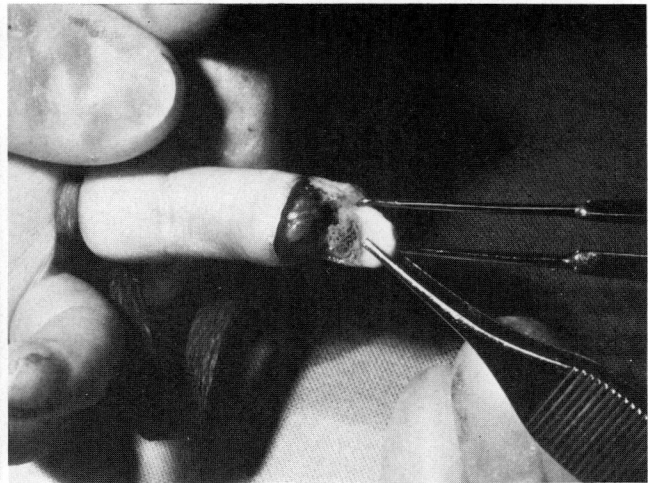

C

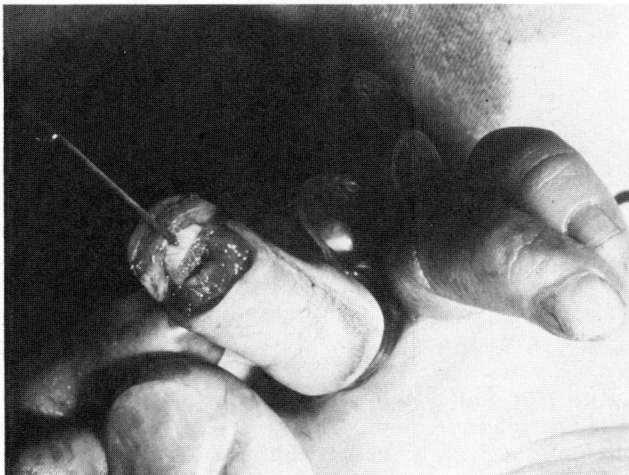

D

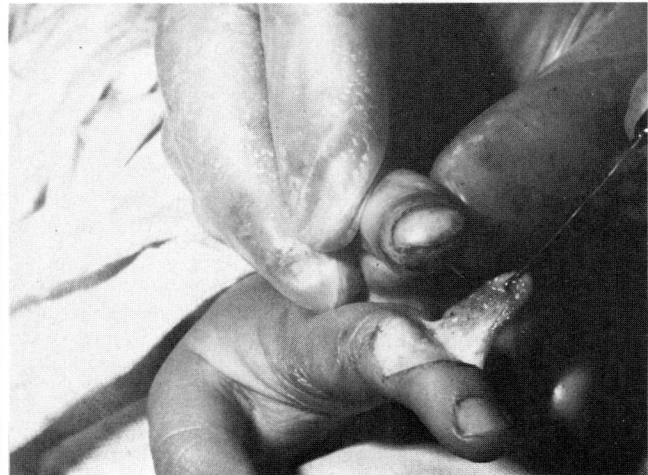

E

Figure 18.4
Patient was 7-year-old girl; (A) deformity as a result of a third degree burn of the distal phalanx on the left hand, middle finger. Partial amputation with loss of nail bed, but the epiphyseal line has been preserved. Claw nail is perpendicular to the long axis of the finger. Finger is lengthened by a semi-circular incision (B) and then iliac cortico cancellous graft is inserted (C); this is re-shaped and secured with a Kirschner wire (D); a cross-finger flap is then applied (E). One and a half years after operation, the new nail is short (F, G, H, I), but the free edge is on the dorsal aspect of the pulp.

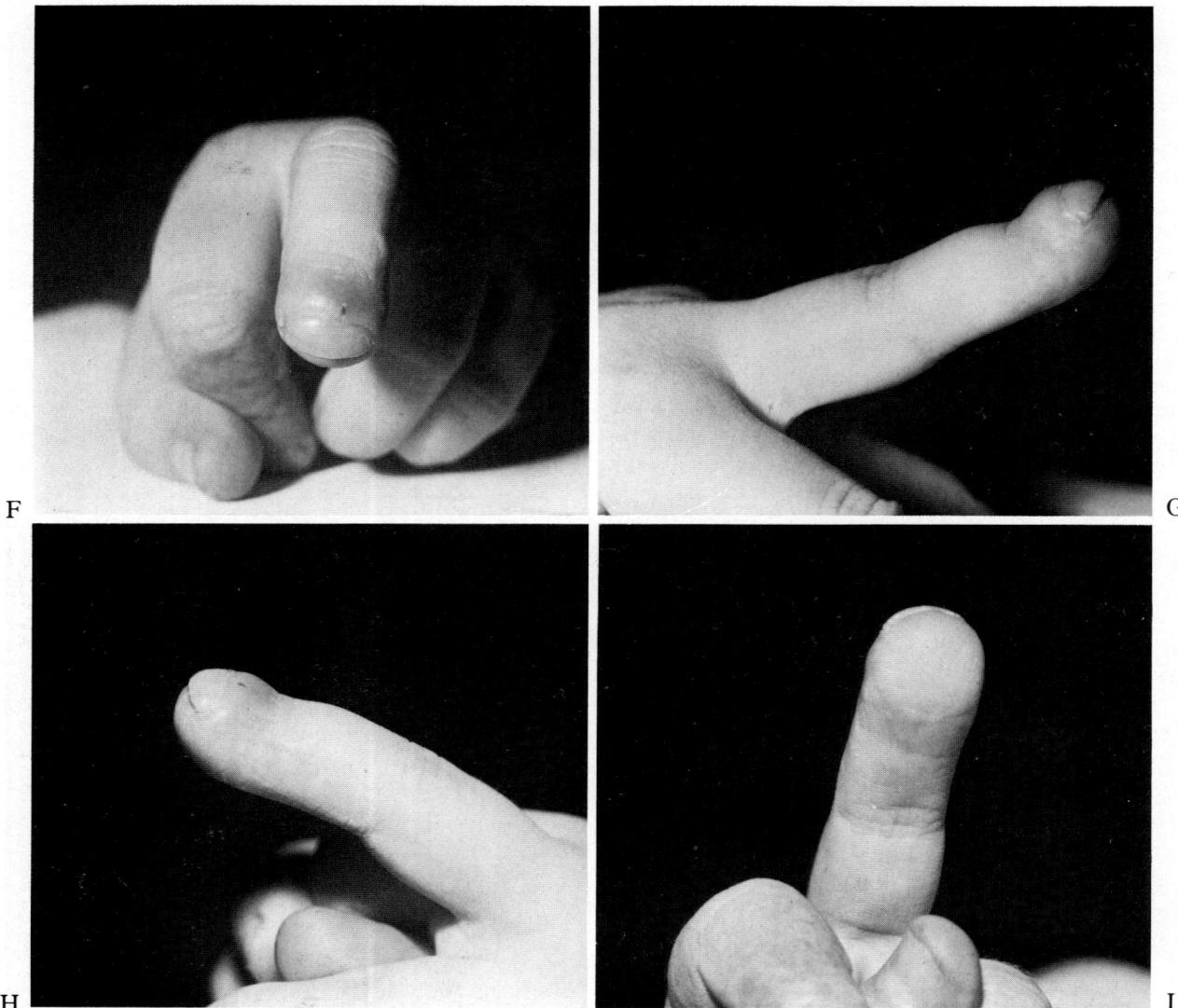

Figure 18.4 (*continued*)

extends by 3–4 mm. At this stage the nail will retain part of its deformity, but later it will tend to flatten on its new support, unless it falls off spontaneously, in which case the new nail that replaces it will be guided to a normal shape.

The palmar aspect of the cancellous bone must now be covered by a skin flap, which is thick enough to act as a new digital pulp and yet stable enough so that it remains secure. The fact that it is fixed to the cancellous bone will help. This flap should not be taken from the inguinal, abdominal or submammary regions as the skin in these parts is insufficiently sensitive and also so thick that it would make un unstable and ugly graft. The two main alternatives are:

(a) a cross-finger flap from an adjoining finger;
(b) a large palmar advancement flap mobilised through mid-lateral incisions.

(a) The first method is fairly reliable as it gives satisfactory cover at least for the index, ring and little fingers, for there is a longer finger than each of these, which can provide the flap. For the middle finger, if the joints are mobile, it is best to use a flap elevated from the base of the thenar eminence. In the case of the thumb, a flap from the dorsal aspect of the index and middle fingers is satisfactory.

The disadvantage of this method is that it does not provide the finger pulp with skin of normal sensitivity in adults and, in addition, it leaves a rectangular defect on the back of the finger. However, this may be covered with a full thickness graft, which will become inconspicuous after a few months.

In children, a flap from another finger or from the thenar eminence provided excellent sensitivity. The author recalls operating many years ago on a blind girl, who was later able to

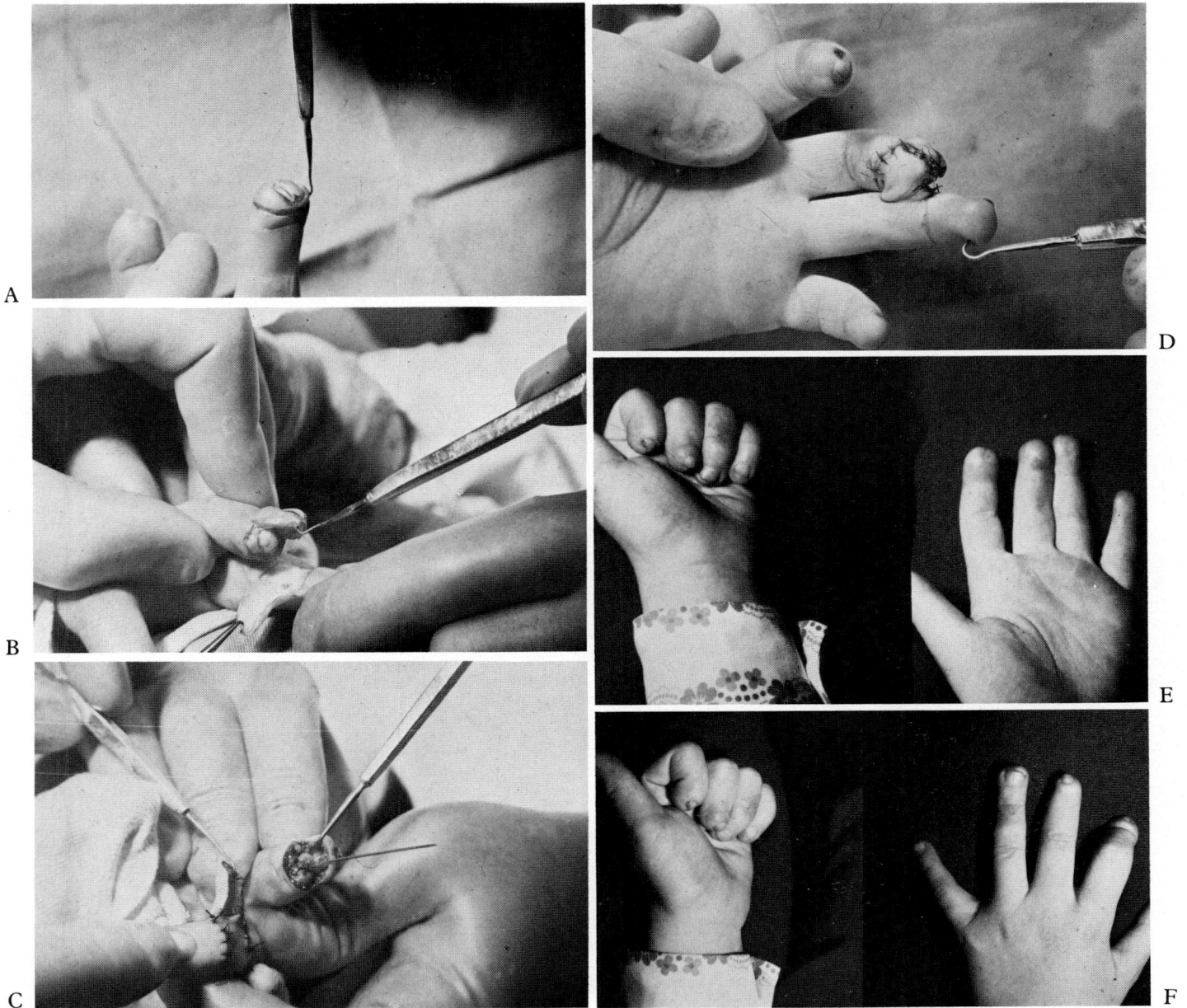

Figure 18.5
Patient was 5-year-old girl. At the age of two, her left hand was caught in a gear. Distal phalanx of middle finger amputated at DIP joint, leaving a small claw nail. Operation in August 1966.
A Semi-circular incision.
B Elevation of the dorsal flap, containing the nail, with exposure of the tip of the middle phalanx.
C Iliac graft in position.
D Cross-finger flap in place.
E and F Result three and a half years after the operation: the nail is small, but correctly positioned; it no longer hinders finger function; the pulp has
 normal sensitivity.

use the finger to learn to read Braille!

Revascularisation of the graft is not a problem (Figs. 18.4 and 18.5).

(b) The second method is a much more delicate procedure. When dissecting the palmar flap, it is vital to cut Cleeland's ligament, taking care to preserve the two neurovascular bundles as far as the base of the finger. The flap is superficial to the fibrous sheath of the flexor tendons, which must not be opened. In the case of the thumb this is relatively easy, for the palmar and dorsal sensory nerves are anatomically well-separated. It becomes much more difficult in the fingers, especially the index and long fingers and the radial half of the ring finger, for the cutaneous sensation of the dorsal aspects of these digits is supplied by branches of the median nerve, arising from the palmar cutaneous nerves. There is a large dorsal branch at the base of the finger, but also fine communications more distally. They cross the mid-lateral incision lines and it is difficult to avoid damaging them, when preparing the palmar flap. Furthermore, the blood supply of the dorsum of the distal and middle segments is derived from fine circular branches coming from the palmar digital arteries. These also will be cut, leaving only the dorsal venous supply and the dorsal arterial vessels, which reach the level of the PIP joint. When the palmar flap is detached, small vessels to the joints will be cut, which also supply the flexor tendons by way of the mesotenon and the vinculae. This may have an adverse effect on the future functioning of the finger. So it is apparent that this method carries more risk than the first. However, it does have the advantage, in theory at least, that it produces a new pulp with perfect sensation. In order to avoid a tender terminal scar, care should be taken to place this as near as possible under the free edge of the nail, where the hyponychium will eventually be. 1–1·5 cm (Fig. 18.3f) can be gained in this way, but the procedure is quite a considerable undertaking; if the lesion is relatively small, the risks of this technique are questionable.

RECESSION OF THE NAIL

The fourth solution uses a different approach: rather than lengthening the finger support, the nail bed and its matrix are brought back in a flap, which is moved rather like a caterpillar's progression. The claw nail must be removed as a preliminary so that the nail bed can adapt itself to its new situation. The length of the stump remains the same and the loss of substance at the end of the finger is replaced by a Tranquilli–Leale plasty (Fig. 18.6).

The success of this technique depends upon retaining the blood supply to the matrix. Normally this is very rich, there being two lateral vessels supplying the lateral margins of the nail structure (Rabischong, 1970). If the nail bed is detached completely from its base, there is a risk of necrosis or dystrophic growth of the nail.

The stability of the regenerating nail depends on its adherence to the keratin of the nail bed. Extreme care must therefore be taken when removing the nail to separate it with a small rugine so

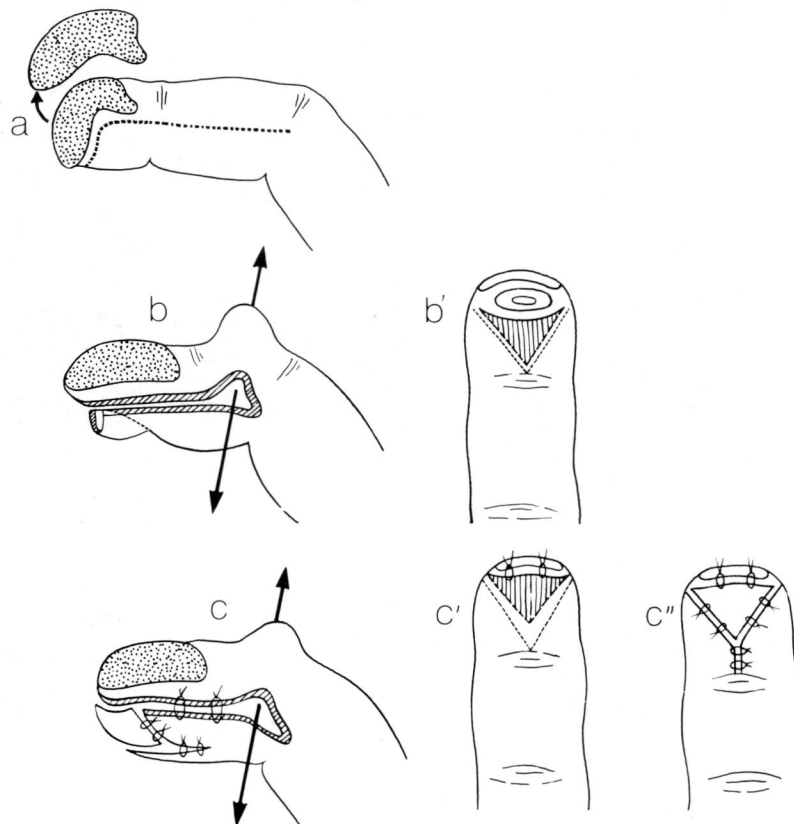

Figure 18.6
Recession of nailbed by caterpillar technique (B and C), combined with an advancement of a triangular flap from the volar aspect in a V–Y plasty B′, C′, C′′.

as to leave the underlying dermal layer intact.

Preservation of the blood supply to the caterpillar flap is all important. After removal of the nail, double dorso-lateral incisions are made and the flap elevated from the paratenon of the extensor tendon, care being taken to preserve the dorsal vein. The nail matrix has to be mobilised from its skeletal anchorages to the terminal phalanx and it follows that the stability of the new nail will be impaired, as its base will now rest near, or on top, of

the extensor tendon. However, since the nail is often incomplete, this matters less than if the nail had been of normal size and shape.

The extent of the recession of the nail is limited to the level of the distal interphalangeal joint, if this remains mobile, but, if this has been stiffened or if the distal phalanx has been completely removed, it may be brought back to the level of the middle phalanx. Figure 18.7 illustrates the extent to which the nail bed can be moved in the case of a boy with arachnodactyly. In this case it was not necessary to remove the excess skin from the caterpillar flap as it reduced spontaneously. Although the nail in its new situation is less stable than normal, it is no longer an embarrassment to normal finger function.

There remains the problem of skin cover to the end of the pulp (Fig. 18.8). This can be done as described in the second method. Having tried various alternative techniques we agree with Simonetta, who has had similar experience, that it is best to close the pulp defect by means of a triangular palmar flap, which is brought forward (Fig. 18.6). This flap is quite sizeable, as it is required to stretch across the width of the terminal segment.

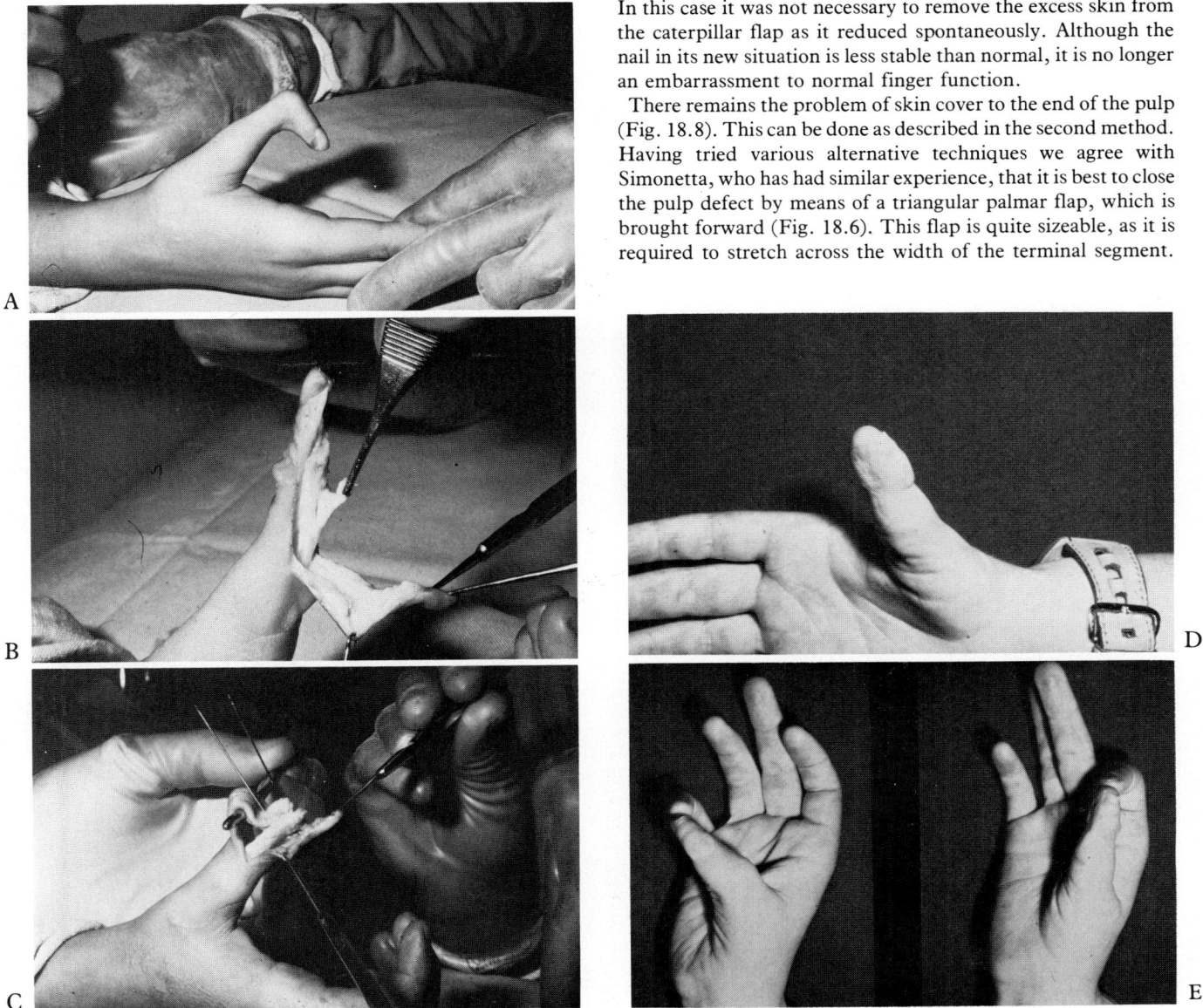

Figure 18.7
Patient was 12-year old boy; arachnodactyly with congenital deformity
A In right thumb; subluxation and curvature of the distal phalanx. Operation in August 1970.
B The terminal phalanx is amputated retaining the flap of palmar skin, which is shortened as required.
C The dorsal flap containing the nail is drawn backwards and attached by caterpillar technique. It is secured by two Kirschner wires through the nail to the bone.
D and E Result at one year: the operation has been successful, both cosmetically and functionally: the nail is stable because the bed has regained attachment to the bone; the dorsal skin fold has reduced spontaneously and no adjustment is required.

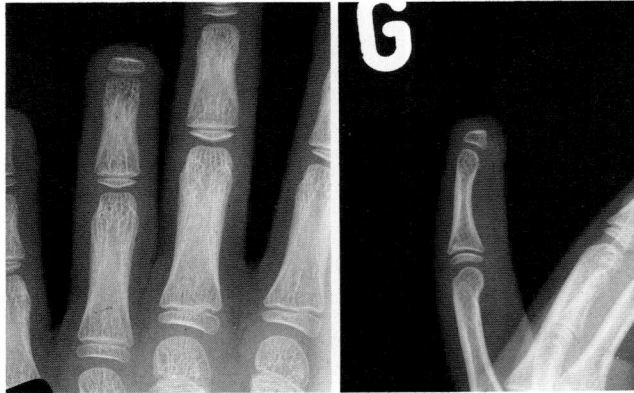

A

Figure 18.8
Patient was a girl, aged nine years; at the age of four, the terminal phalanx of the left ring finger was amputated in a door; treated by means of a partial skin graft. Claw nail has developed.
A X-ray.
B Mid-lateral incisions made on both sides plus V-incision of the pulp (Tranquilli–Leale's type) forming one continuous strip.
C and D Triangle of pulp is anchored to bone with fine Kirschner wire.
E and F The result at five months.

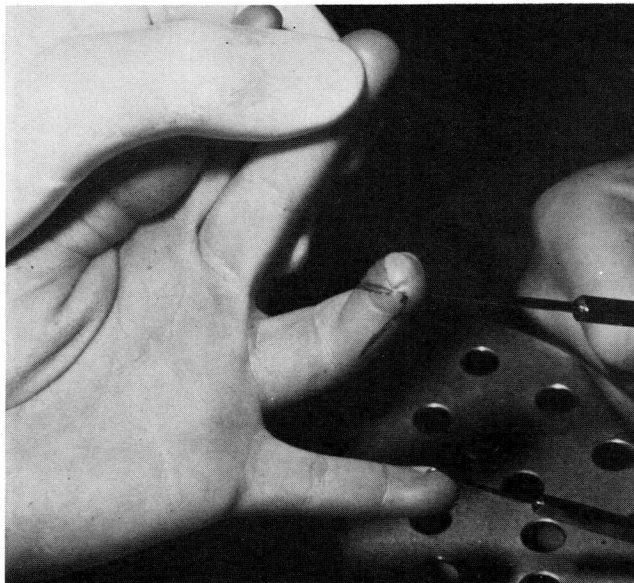

B

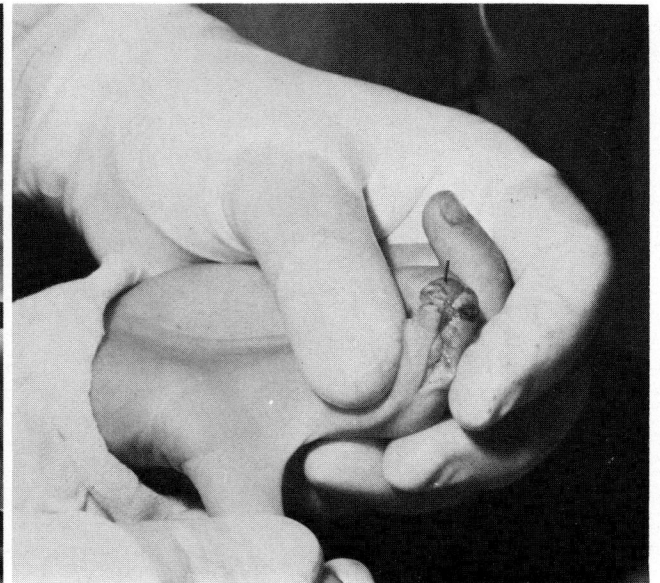

C

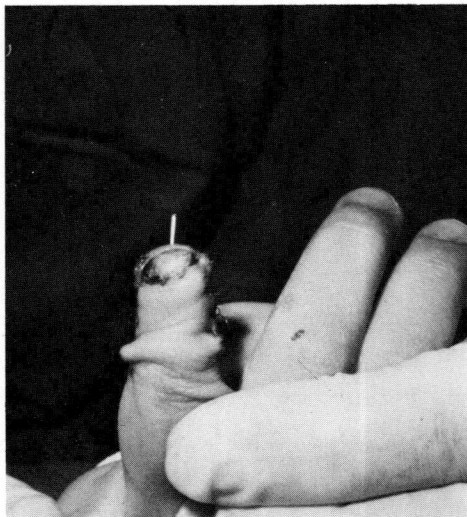

D

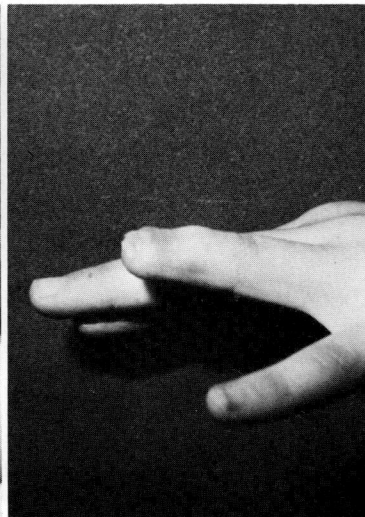

E

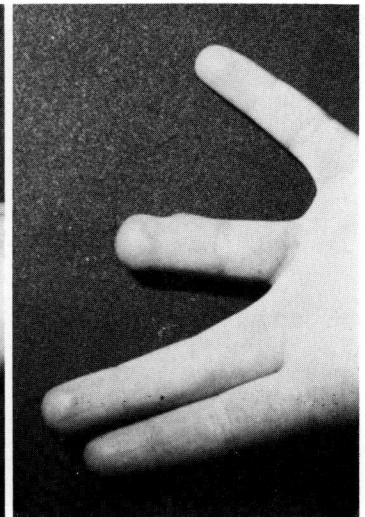

F

Care must be taken to preserve the terminal branches of the two digital nerves, which can be clearly seen under magnification. The artery and its terminal branches are also visible, although deep to the nerve. The delicate fibrous strands, which partition the pulp and attach it to the sheath of the flexor tendon and to the periosteum, are cut with fine scissors. A sort of bipedicle strip is thus formed, which must slide without strain to cover the end of the bone, and the flap may then be secured by a fine Kirschner wire to the bone end. The wound is then closed in a Y-shape, using 5–0 nylon. It is important that tension should be avoided at the suture line, otherwise the flap may become avascular. A bulky dressing will ensure adequate immobilisation without the use of a splint, as the latter might exert pressure on the precarious flaps. If the Kirschner wire appears to cause ischaemia, it should be immediately removed once the stitches have been inserted.

Figures 18.6a–18.6c illustrate the technique, which has the great advantage of requiring no extra tissue apart from that which may be obtained from the finger itself.

CONCLUSION

Although in some respects the claw nail appears a trivial condition and is consequently often neglected, it can be a source of significant disability. It is possible, although difficult, to treat it surgically by the plastic techniques, which have been described; these can give good results, although the success may be limited by the nature of the original injury.

BIBLIOGRAPHY

[1] RABISCHONG P. — Communication faite à la Réunion des 5, 6 juin 1970 du GEM à Marseille.
[2] SIMONETTA C. — *Chirurgie réparatrice des pertes substance et des amputations du pouce.* Thèse Méd., Université de Lausanne, 1975.
[3] VERDAN Cl. — Chirurgie des lésions traumatiques récentes de la main. *Helv. Chir. Acta*, 1956, *23*, 411–451.

19. MUCOUS PSEUDOCYSTS OF THE FINGER

E. Moberg

In the literature, these cysts are nearly always described as being degenerative, the result of mucoid transformation or metaplasia of the connective tissues of the chorion.

Kleinert, Kutz, Fishman and McGraw (1972), however, have shown that it is possible in some cases to dissect a track running from the cyst to the joint capsule. As Andren and Eiken (1971) have demonstrated, synovial cysts of the wrist are indeed often formed from synovial fluid released under pressure, sometimes from arthritic joints.

They discovered that fine, synovial tracks originating in the joint often led to these cysts. Using arthography, the cyst could be outlined by injection of contrast medium under pressure. Microscopic examination revealed in several cases that the joint capsule had deep narrow recesses, which the authors considered normal, but which could, however, be regarded as pre-cystic structures. Synovial fluid, pumped out of the joint by movement, distended these recesses. Valvular mechanism prevented the fluid from returning to the joint and the contents of the cysts subsequently became more concentrated and mucoid.

Eiken has allowed me to report that he has made similar observations in relation to cysts of the fingers and I have been able to fill such cysts with a blue dye injected under pressure into the distal joint. I feel justified, therefore, in considering these cysts to be synovial cysts originating from the joint (Fig. 19.1).

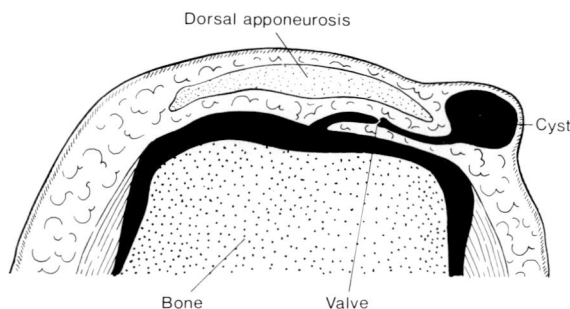

Figure 19.1
A mucous pseudocyst with its track and valve.

A valvular mechanism appears to be a pre-requisite for their development. These cysts, therefore, have the same pathogenesis as the synovial cysts of the wrists, Baker's cysts, the rare large cysts on the anterior aspect of the hip joint, and the cysts of the peroneal nerve, which originate in the superior tibio-fibular joint, and which have been so well demonstrated by Athol Parkes (1961). It is very important to recognise this from the point of view of treatment. These cysts appear on the dorsal surface of the terminal phalanx (Fig. 19.2), nearly always in the

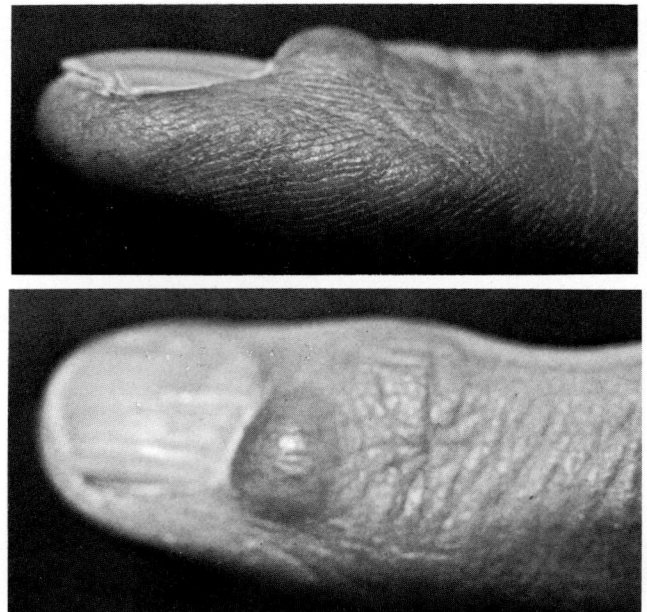

Figure 19.2
Cyst on the finger.

middle-aged or elderly, and often in association with an arthritic joint. About 75 per cent of the cases occur in women. Cysts may appear on several fingers. There are no signs of inflammation around an intact cyst. The overlying skin is often extremely thin and transparent and usually the cysts have burst on at least one occasion, releasing gelatinous content and healing, followed by a recurrence a few months later. In some cases the cysts produce pressure on the nail matrix, resulting in longitudinal grooves on the nail. As the cysts develop lateral to the extensor tendon, these grooves also are laterally placed in the nail.

The cysts are generally painless, but can, however, become sensitive and mildly inflamed if they break down. They are rare in manual workers, but, if they do occur, treatment is important. Most commonly the patient is concerned about the appearance of the cyst, but declines treatment when she learns that the lesion is harmless.

The evolution of these lesions is variable: some cases disappear without treatment, while others disappear following mild infection, and some appear to be cured by aspiration and hydrocortisone injection. In these latter cases, it is probably the irritation caused by the solvent rather than the hydrocortisone itself which exerts the benefit. Many other methods have been used but are not worth discussion.

Surgical excision produces a definitive cure in two-thirds of cases. However, recurrence may occur more than a year after excision and results are less encouraging than might be expected.

Enucleation of the cyst is seldom possible because of the thin adherent skin covering. It is rarely possible to close the skin after removal of the cyst by primary suture. A small flap of adjacent skin may be used, or, if necessary, the site of excision may be covered by a free skin graft. The high failure rate of surgical treatment is understandable in the light of the pathogenesis of these cysts described above.

Only Kleinert and his colleagues have produced healing in all cases by means of excision and dissection of the track. This more radical operation demands a wider exposure so that the pedicle may be traced to the joint and a window must be made in the capsule. Closure of the wound may then be a problem and there is always a risk that pathogenic organisms, lingering in the nail, may complicate healing. This delicate surgery is, therefore, not free from risk, particularly of septic arthritis of the distal joint. The stitches should not be removed early and the joint should not be mobilised until the soft tissues are soundly healed.

If it is agreed that these are cysts of synovial origin, then it should be possible to effect a cure by excising the part of the capsule from which the pedicle arises, thus destroying the valvular mechanism, which would allow the cyst to disappear spontaneously. This procedure should avoid complications and should make skin closure easier, but as far as I am aware this simple technique has not been tried.

BIBLIOGRAPHY

[1] ANDREN L., EIKEN O. — Arthrographic studies of wrist ganglions. *J. Bone Jt Surg.*, 1971, 53 A, n° 2.

[2] ARNER O., LINDHOLM A., ROMANUS R. — Mucous cysts of the fingers. Report of 26 cases. *Acta chir. scand.*, 1956, 111, n° 4.

[3] EIKEN O. — Personnal communication.

[4] KLEINERT H. E., KUTZ J. E., FISHMAN J. H., McGRAW L. H. — Etiology and treatment of so called mucous cysts of the finger. *J. Bone Jt Surg.*, 1972, 54 A, n° 7.

[5] PARKES A. — Intraneural ganglion of the lateral popliteal nerve. *J. Bone Jt Surg.*, 1961, 43B, 784–790.

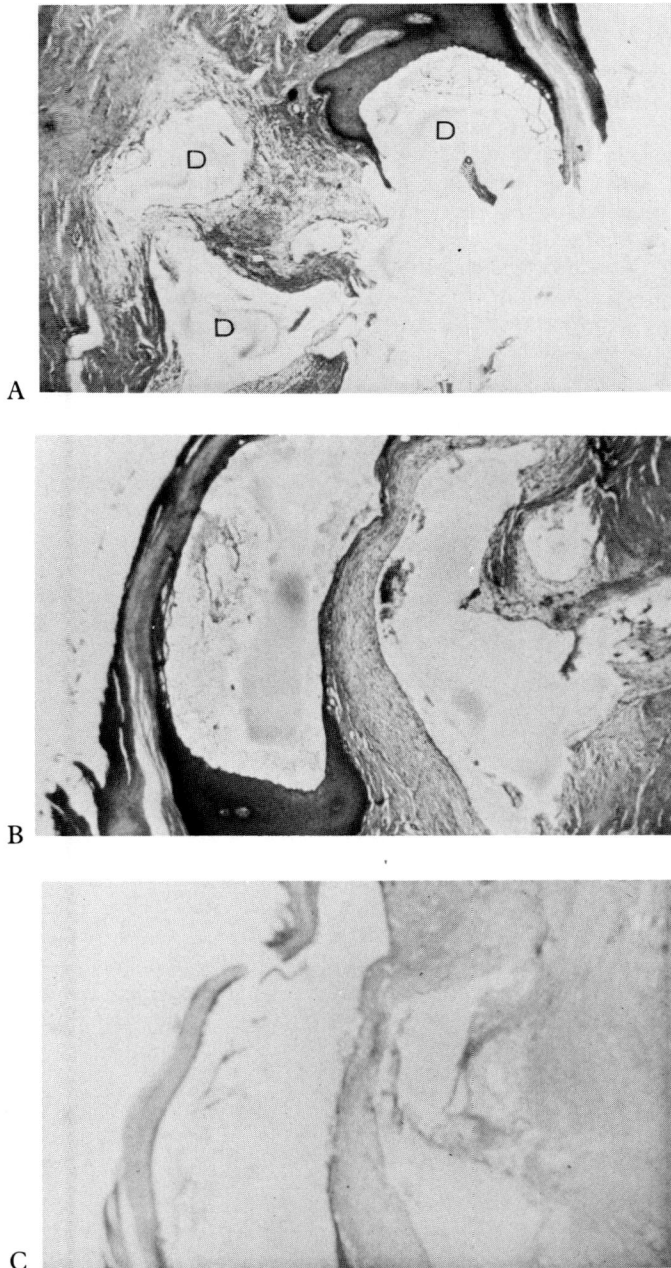

Figure 19.3
A Cutaneous surface of the pseudocyst.
B The sinus track.
C. Communication of the tract wth the joint.

20. FREE NAIL GRAFTING

C. R. McCash

The successful grafting of a finger- or toe-nail along with its germinal matrix to another digit is not specially difficult for any surgeon familiar with the technique of Wolfe grafting. The fact that it is not often attempted is mainly because, although the graft may survive and grow again in its new position, one cannot be sure that the new nail will remain cosmetically acceptable.

Some 20 years ago I carried out an experimental series of 25 free nail grafts on cases of congenital absence, traumatic loss, and destruction by burns. In 22 of these, recognisable new nails grew again, but only five could be considered normal in appearance. Using the technique of transfixion with a wire mattress suture (Fig. 20.1) and careful selection of suitable cases, a higher proportion of successful results can be expected.

The interesting anatomical studies of Professor Achten on the growth of the nail in three layers may well explain some of the partial failure which occurred. From my experience the criteria for a successful nail graft are as follows:

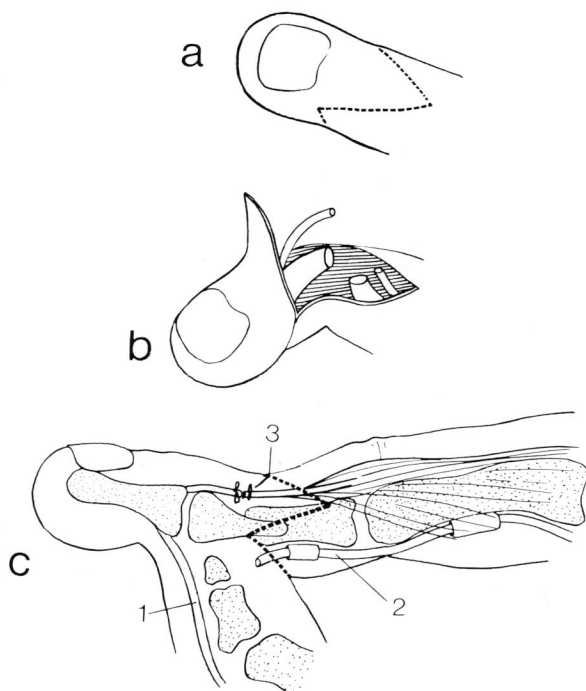

Figure 20.1
Technique for toe–finger transplantation.
a. Incision.
b. Division of the extensor tendon and middle phalanx.
c. Transfer: 1. long flexor tendon of the toe; 2. long flexor tendon of the finger; 3. suture of the extensor tendons.

(a) the donor nail must be suitable in size and texture
(b) the recipient nail bed must be vascular enough to nourish the graft
(c) the upper surface of the phalanx upon which the graft will rest must be smooth and contoured if necessary, so that it fits accurately the under surface of the new nail
(d) following the operation the graft must be held firmly pressed against the phalanx for two weeks to prevent it being detached by a haematoma
(e) the patient must understand that it will be his responsibility to protect the nail graft from injury or infection for the first 4–6 weeks.

TWO STAGE TOE-TO-FINGER TRANSFER

This is the one reliable way of replacing a nail loss in cases where it is justified. Such cases are usually traumatic where there is also loss of length in the digit affected. It is a much more complicated procedure for both patient and surgeon. It involves the transfer in two stages of the distal half of a toe with one and a half phalanges and requires the foot and hand to be joined together for not less than two weeks (Figs. 20.2, 20.3 and 20.4). For the thumb, the great toe would be used (Clarkson, 1966) and for a finger the second or third toe.

The advantage this method has over free nail grafting is that there is no interference with either nail or nail bed, and after a time, the colour and texture of the toe-nail grows to look more like a finger-nail.

The recovery of tactile sensation in the grafted portion is very rapid and there is no necessity to suture the digital nerve even if this were possible. In three cases the results have been satisfactory for patients and surgeon. Two were to the middle finger and one to the index. Nine years after operation, one case followed up still showed a permanent good result. She was a married woman with two children and used the hand for all normal domestic duties.

BIBLIOGRAPHY

[1] CLARKSON P. (1966). Toe Hand Transfers. *Flynn Hand Surgery*, 583.
[2] PAPAVASSILIOU N. P. (1969). Transplantation of the Nail. *Brit. J. Plas. Surg.* **XXII**, 3, 274.

Acknowledgement of permission to use photographs and diagrams which previously appeared in *Transactions of Third International Congress*, 1963 Series 66 pp. 977–79 given by Excerpta Medical Foundation, Amsterdam, the publishers.

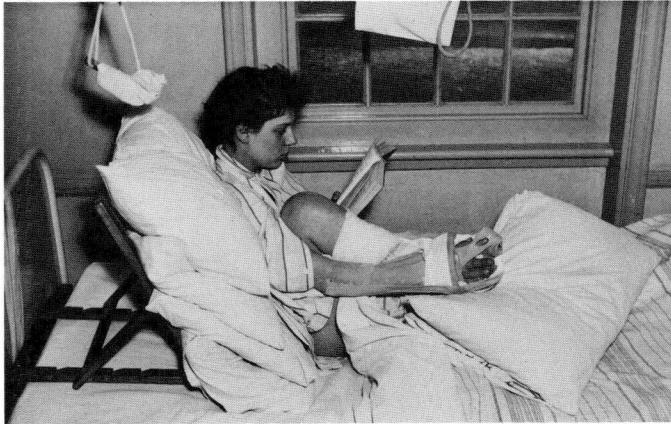

A

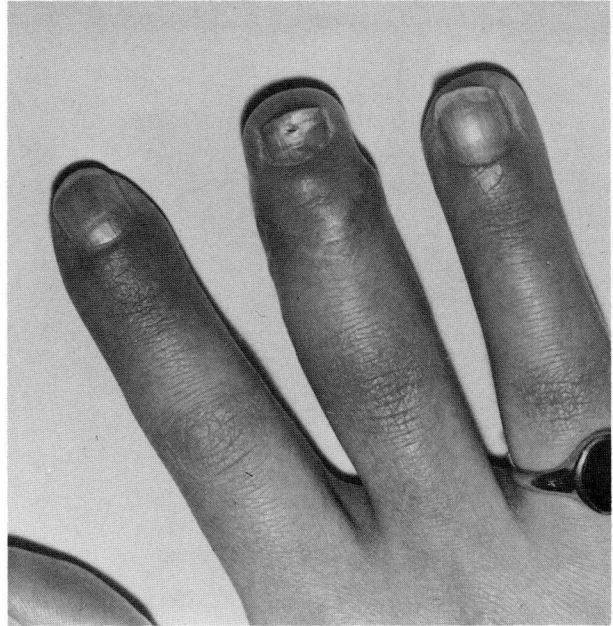

B

C

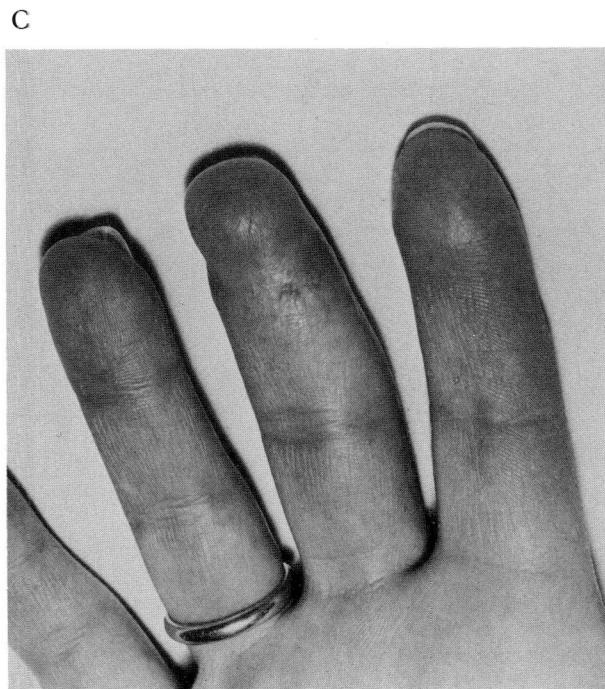

Figure 20.2
Transplantation of a toe-nail to a finger. The injured finger was the middle finger of the right hand. Distal phalanx was severed by a cutting machine in 1955, when the patient was 17 years old. Reconstruction took place in 1956.

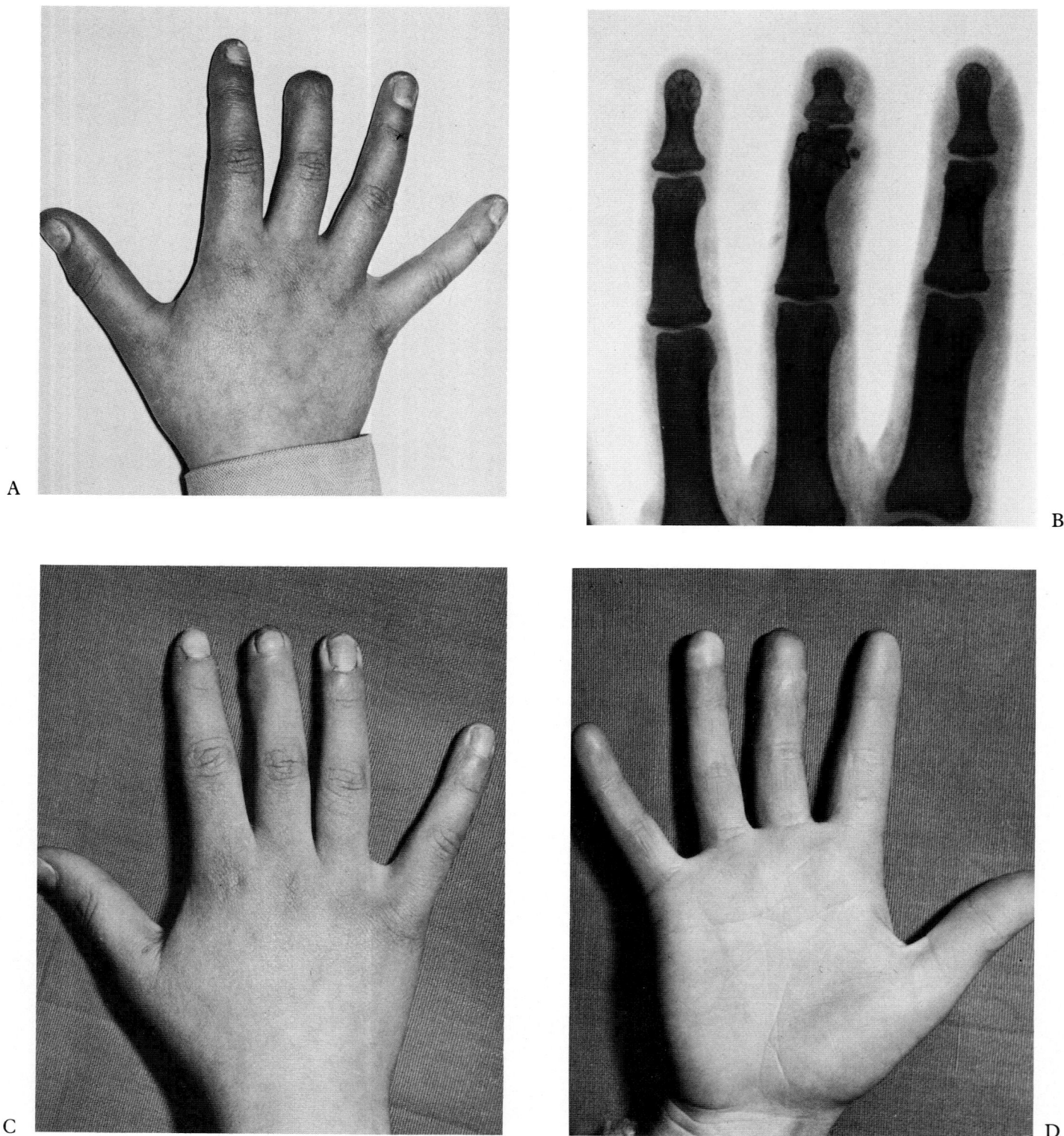

Figure 20.3
Transplantation of a toe-nail to a finger. The patient was a young girl aged 15 years, who had lost her fingertip at age of two. Duration of hand–foot fixation before separation of graft was two weeks.

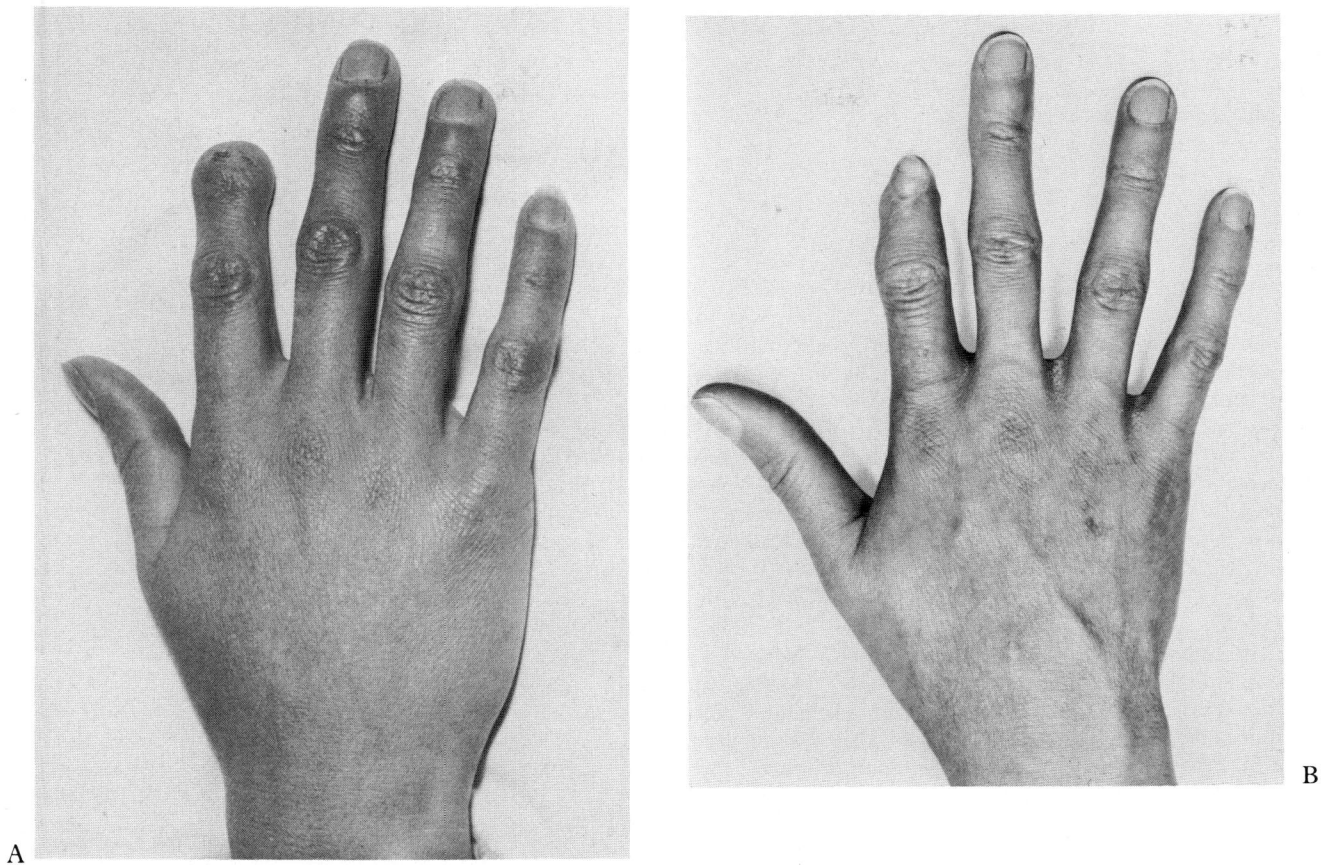

Figure 20.4
Transplantation of a toe-nail to a finger carried out in 1958. In this case, part of the bone of the toe was lost, thereby shortening the finger. Three years later the finger had not changed in appearance, the nail was growing normally and the patient was satisfied.

21. RECONSTRUCTION OF THE FINGER-NAIL BY MICROVASCULAR TRANSFER FROM THE TOES

W. A. Morrison

Apart from its obvious aesthetic value the nail, whose distant ancestor is the claw, still retains important elements of that function. Defence has given way to refined grasp and this utilises two important components of nail anatomy. Because of the rich nerve supply at the nail bed it acts as a sensitive organ of touch which is amplified by transmission along the nail length in a similar fashion to that of the cat's whiskers. It also has an important function as longitudinal and lateral extraskeletal support for counter-pressure to the pulp. The resultant fine distortion of the nail bed by pressure magnifies the tactile appreciation of pinch (Rank *et al.*, 1973).

The thumb is the most important digit especially for pinch and as such relies heavily on its nail for counter-pressure, sensory magnification and delicate pick-up manoeuvres.

Absence of the nail is one of the obvious cosmetic and functional disabilities of a conventionally reconstructed thumb. Digital transposition and toe-to-thumb transfer overcome this problem but they are indicated only for major ablations.

Fortunately the big toe-nail size approximates to the thumb closely, some patients more so than others and this lends itself to distant transfer by microvascular anastomosis because of the large arterial and venous anatomy of the big toe. The vessels are not specific to the nail and its bed, however, and a proportion of the toe pulp must also be transferred to assure vascularity of the transfer.

Most hand injuries involving the nail and its bed also involve the terminal phalanx to a greater or lesser degree so that a composite of the nail and toe pulp is usually advantageous. Such procedures are ideally suited to the complete or partial reconstruction of the thumb where the nail is big and comparable to the toe and glabrous innervated pulp is also required to reconstitute the vital pinch mechanism.

Lesser toe-nails are usually too small to look convincing for finger-nail reconstruction although occasionally this is not so (Fig. 21.1).

Transfer of toe-nail and bed for the sole purpose of nail reconstruction without pulp has not been done in this series. For finger-nail reconstruction as distinct from thumb, it is theoretically feasible to transfer two-thirds of the big toe-nail only in order to match the finger-nail width more accurately although this also has not been done.

TECHNIQUE

The value of any transfer procedure is to gain as much as possible with minimal secondary deformity. The nail, with its germinal and non-germinal bed and an underlying sliver of

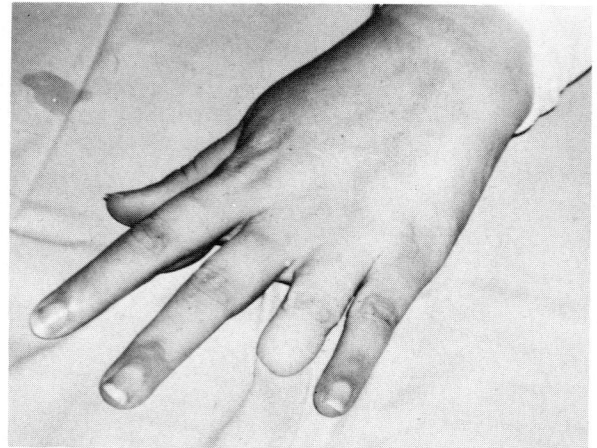

A

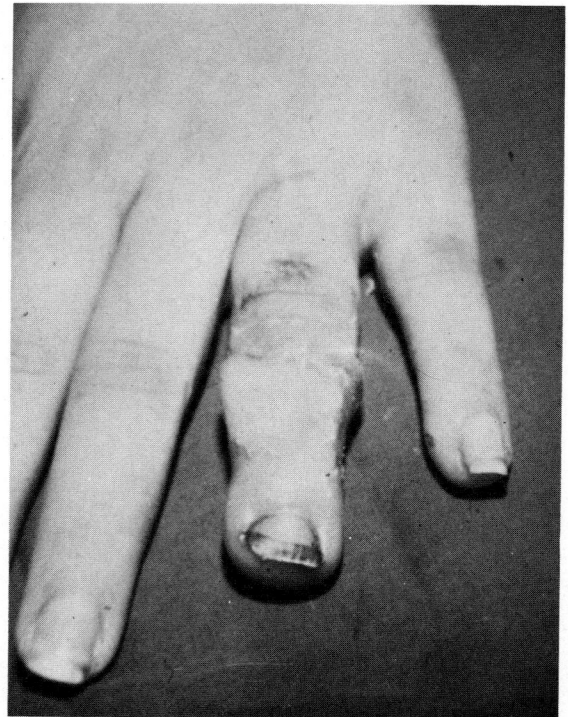

B

Figure 21.1
A 40-year-old woman with an amputated left ring finger at distal mid-phalangeal level requested reconstruction for cosmetic reasons.
B 2 months after transplantation of distal two phalanges of the second toe (reproduced with kind permission of Mr B. O'Brien).

terminal phalanx for stability, can be elevated with a greater or smaller portion of the lateral pulp and the proximal dorsal skin depending on individual needs.

ANATOMY

The predominant arterial supply is usually from an extension of the dorsalis pedis system via the first dorsal intermetatarsal artery. This provides an easily anastomosable artery of long length and large diameter. When this dorsal system is small the plantar one is reciprocally large and again can be readily used for transfer to the hand. The venous drainage arises from the dorsal aspect of the proximal nail bed flap and via a rapid confluence forms the dorsal venous arch system becoming the long saphenous vein. These veins are also long and of large diameter making transfer easy although at their origin they are extremely fine and require conservation of some surrounding subcutaneous tissue in order to prevent damage. The nerve supply to the lateral pulp is twofold, predominantly from the medial plantar nerve but also dorsally from the terminal branches of the deep peroneal nerve.

SECONDARY DEFECT OF THE TOE

Pilfering of the toe hemipulp and its nail creates very little functional or cosmetic secondary disability. The pulp defect readily accepts a split thickness skin graft and the bared portion of bone underneath the nail is ideally managed by a cross-toe flap from the volar aspect of the adjacent second toe (Fig. 21.2).

RECIPIENT SITE

The toe segment is removed in such a manner as to fit the recipient site. However, because the nail bed requires a protective proximal flap the recipient defect sometimes has to be tailored to accommodate the flap. Ideally some underlying terminal phalanx of the toe is preserved beneath the germinal

part of the nail bed to assure uninterrupted growth and this again may have to be accommodated at the recipient site. Usually sufficient length of artery and vein are obtained from the foot to allow easy anastomosis to the recipient ulnar digital artery of the thumb at the proximal first web space or even more proximally to

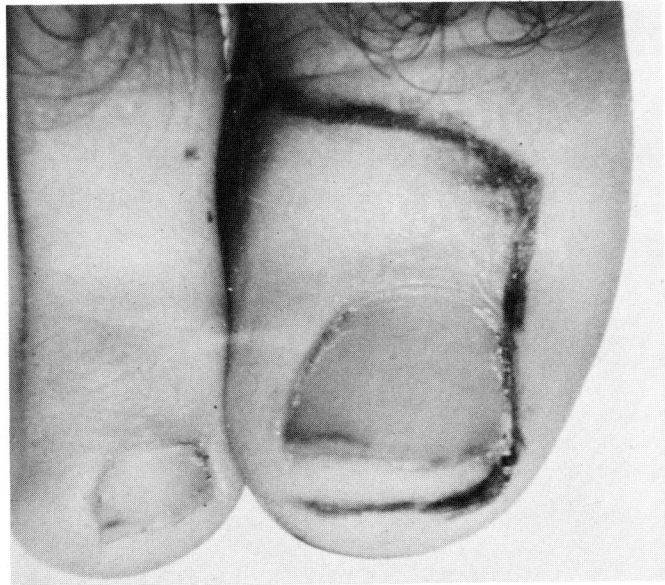

B

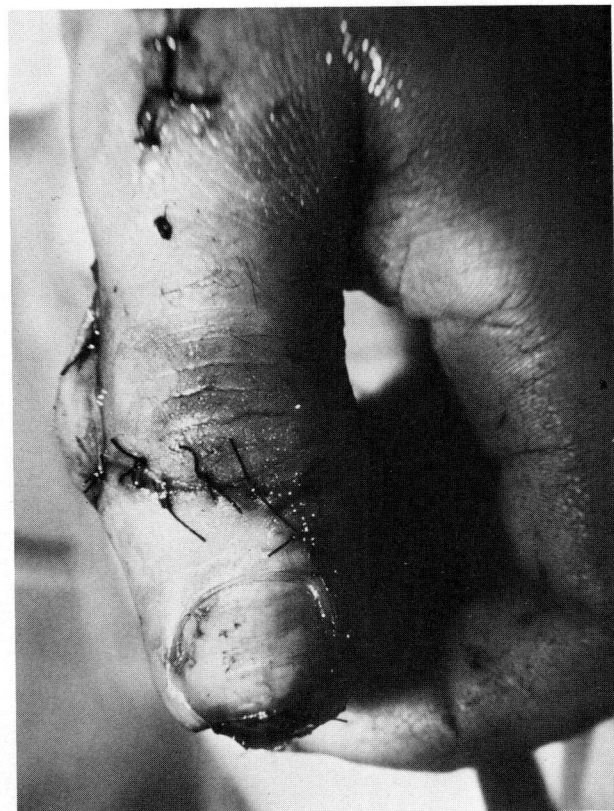

C

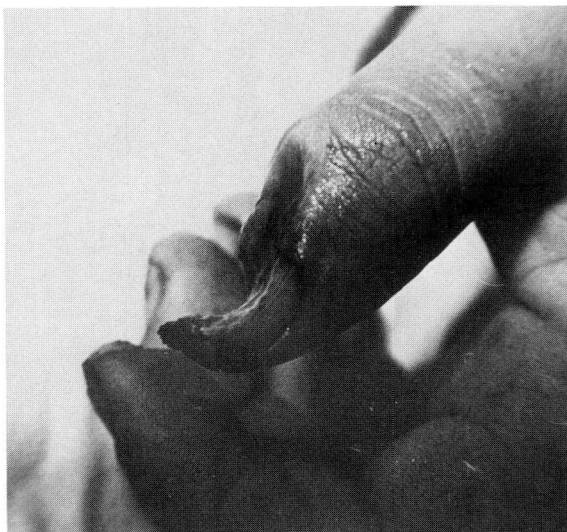

A

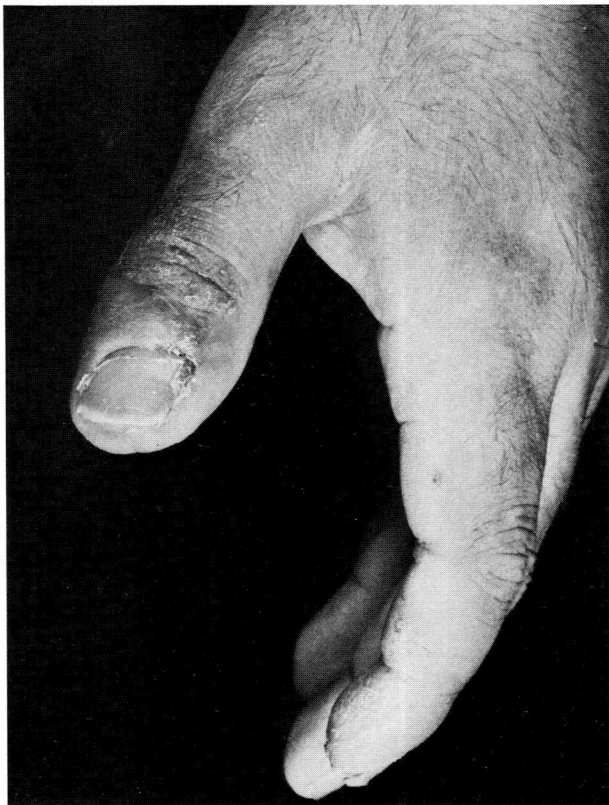

D

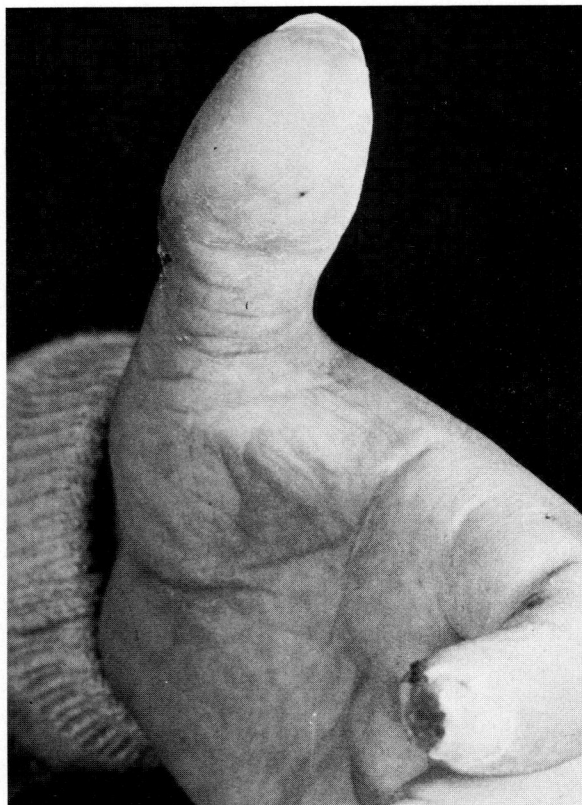

E

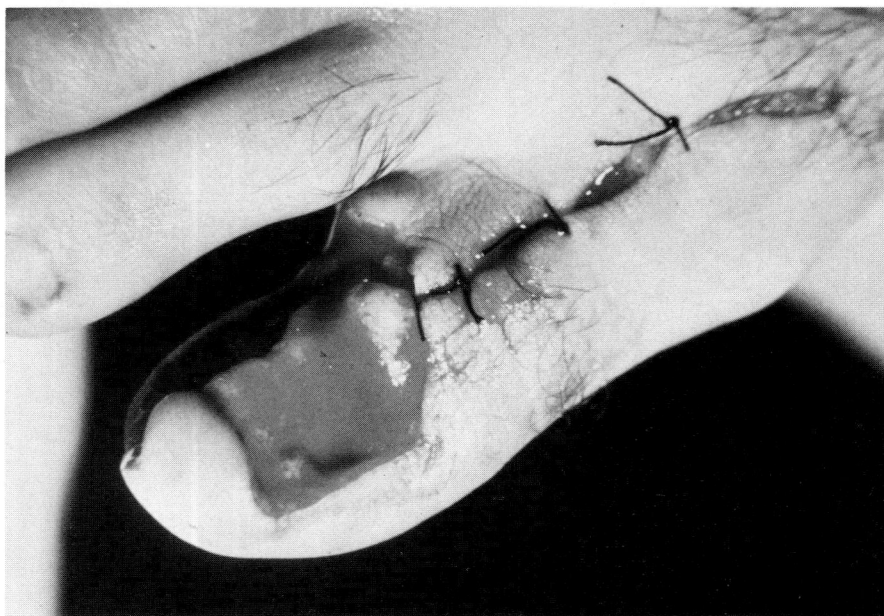

F

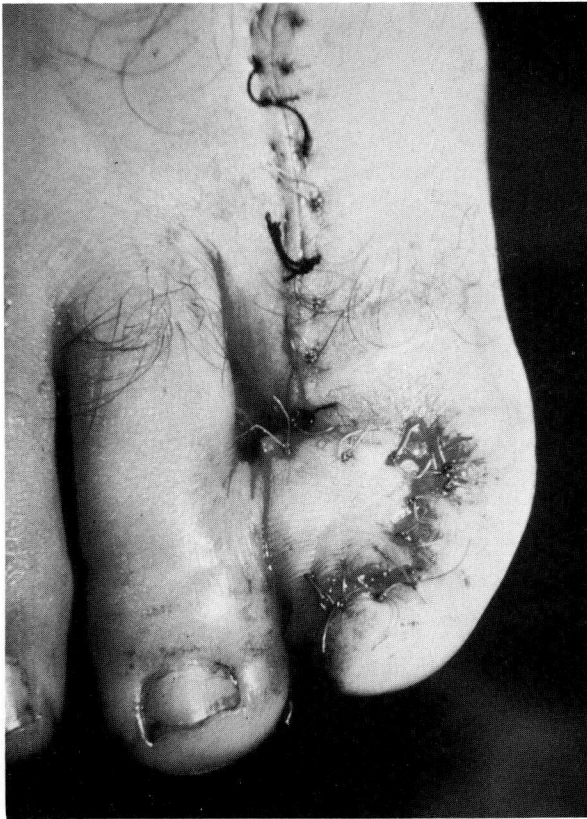

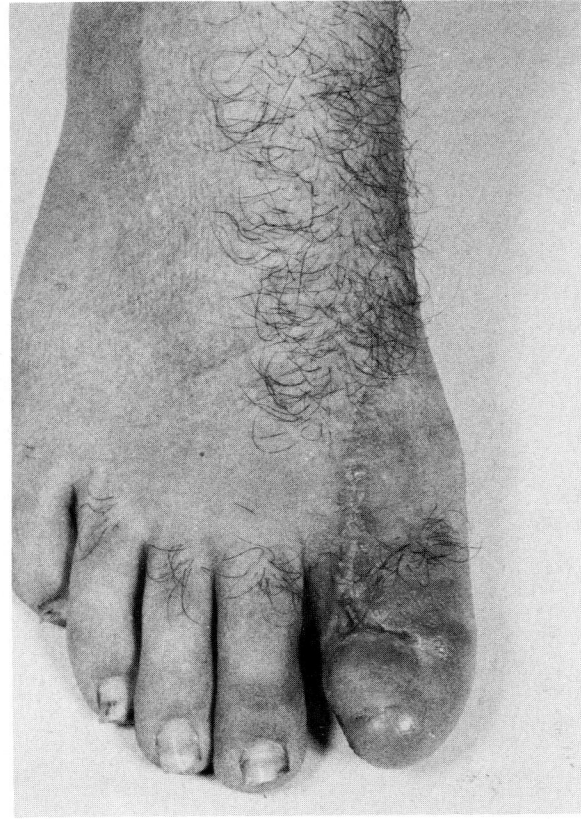

Fig. 21.2
A Amputation of radial two-thirds of distal phalanx left thumb in 25-year-old left-handed carpenter.
B Proposed segment of right big toe for transfer to match thumb defect.
C Immediate post-operative appearance after anastomosis of plantar digital nerve and artery to the proximal radial digital nerve and artery of the thumb and the long saphenous vein to cephalic vein.
D 3 months post-operatively dorsal view.
E 3 months post-operatively volar view.
F Donor defect prior to closure.
G Cross-toe flap to bone defect.
H Donor defect at 3 months.

the radial artery in the anatomical snuff box. The vein is anastomosed to one of the multitudinous dorsal veins in this region. Both flap nerves are joined to the ulnar digital nerve.

POST-OPERATIVE MANAGEMENT

Heparin is not used but asprin 300mg T.D.S. and Persantin 25mg Q.I.D. are given for the duration of the patient's stay in hospital.

LONG-TERM RESULTS

All four cases in the series have survived. The longest follow-up period is eighteen months (Fig. 21.3) and in this case some underlying terminal phalanx was preserved. After initial loss of the nail the new nail regrew in a normal fashion. More recently

cases have been performed where no underlying bone was taken but simply elevating the nail bed in a subperiosteal plane (Fig. 21.4). These patients again initially shed the nail but a new nail of normal appearance is growing. The long-term quality from the point of view of appearance and stability is not yet fully known. Microvascular surgery affords a safe method of transfer of toe-nails to the hand for nail reconstruction. It is particularly suited to thumb reconstruction where a combination of innervated pulp and nail is required.

BIBLIOGRAPHY

[1] RANK B. K., WAKEFIELD A. R. and HUESTON J. T. — (1973). *Surgery of repair as applied to hand injuries*, 4th Edition, pp. 19 and 20. Edinburgh: Churchill Livingstone.

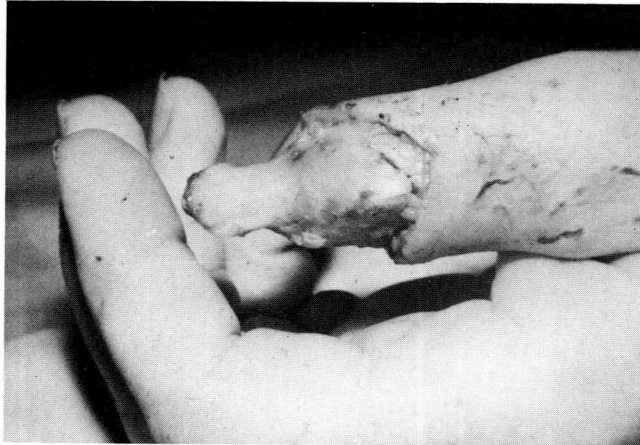

A

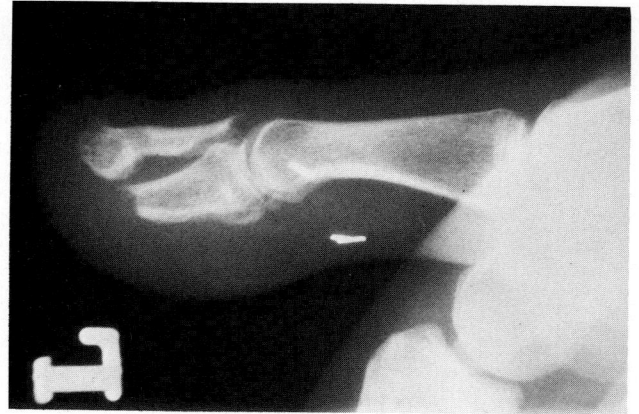

D

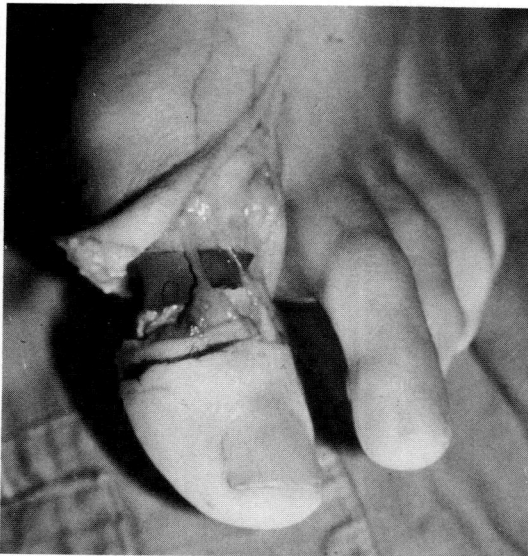

B

C

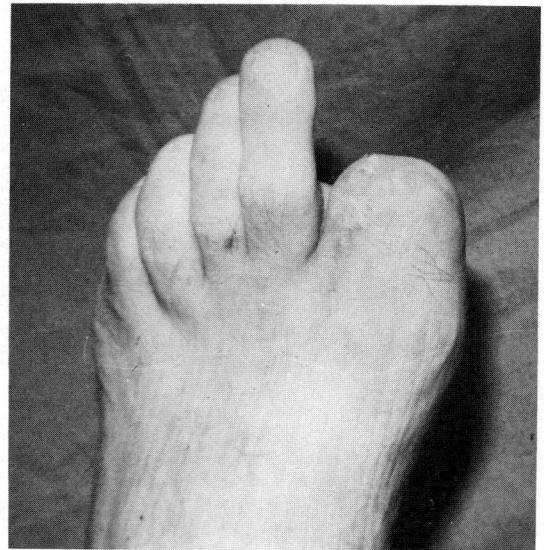

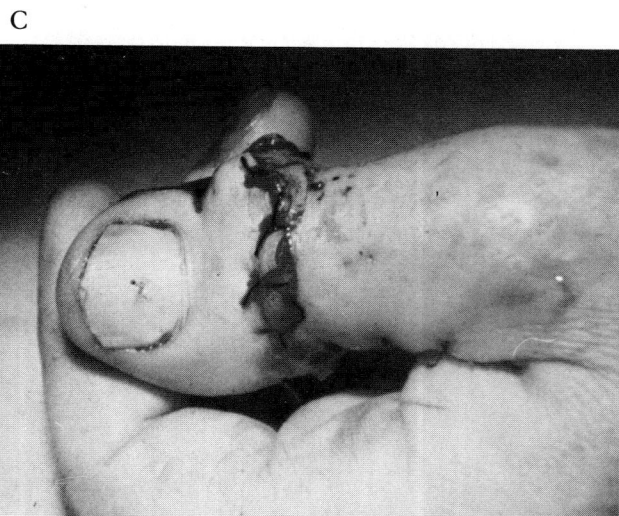

E

Figure 21.3
A 45-year-old man with a degloving injury of the left thumb preserving
   distal phalanx and tendons.
B Donor flap including nail removed as a cap showing dorsal venous
   system.
C Toe segment capped over the degloved thumb distal phalanx.
D X-ray showing preservation of sliver of terminal phalanx of toe to
   protect the overlying nail growth plate.
E Donor defect showing toe amputated just proximal to the I.P. joint.

Figure 21.4   ▶
A 19-year-old man with an amputated left thumb at proximal phalangeal
   level. He sustained considerable damage to his fingers as well.
B Dissection of dorsal and lateral skin of the ipsilateral big toe to be
   used as a one-stage complete wrap around from iliac bone graft
   for thumb reconstruction.
C Dissection showing the dorsal intermatatarsal artery.
D Result at four months.

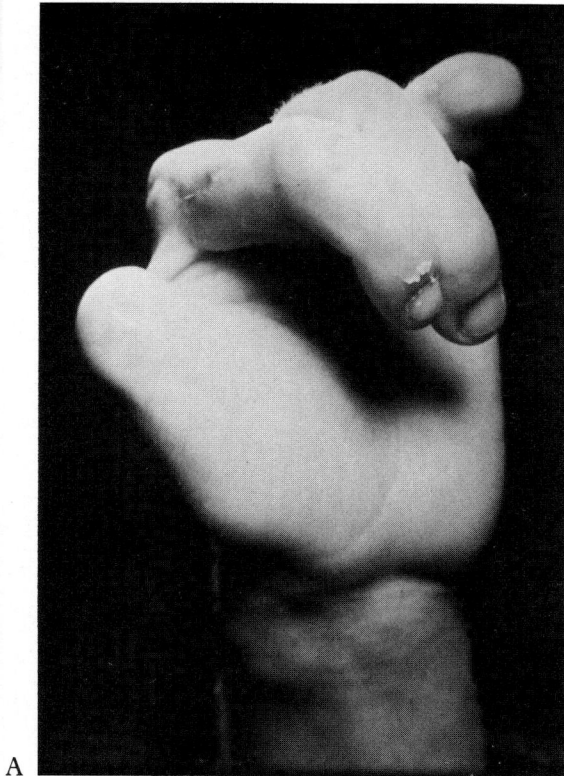

A

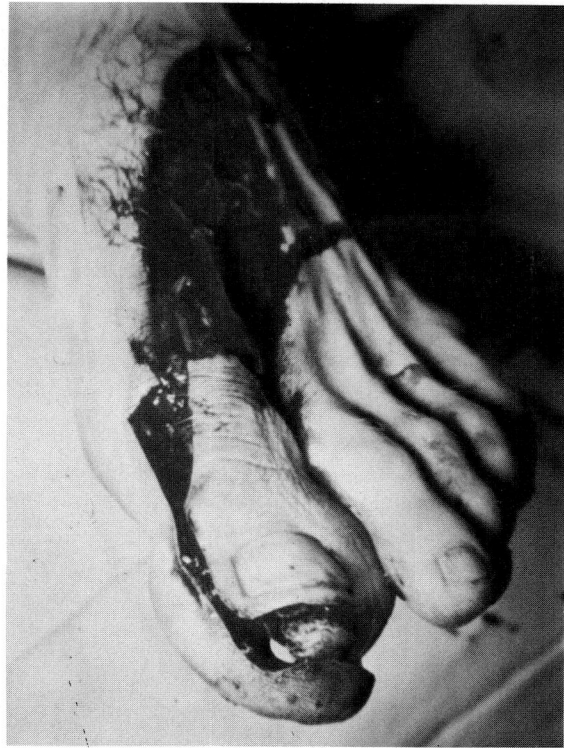

B

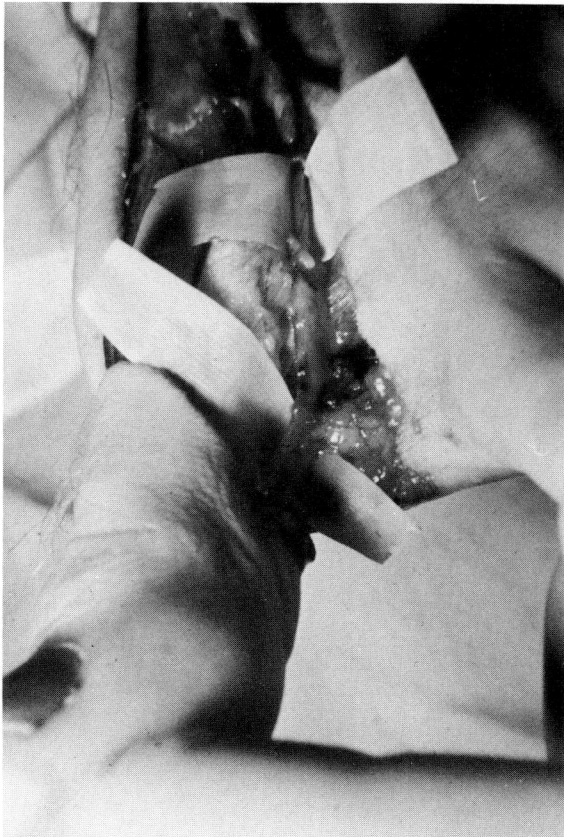

C

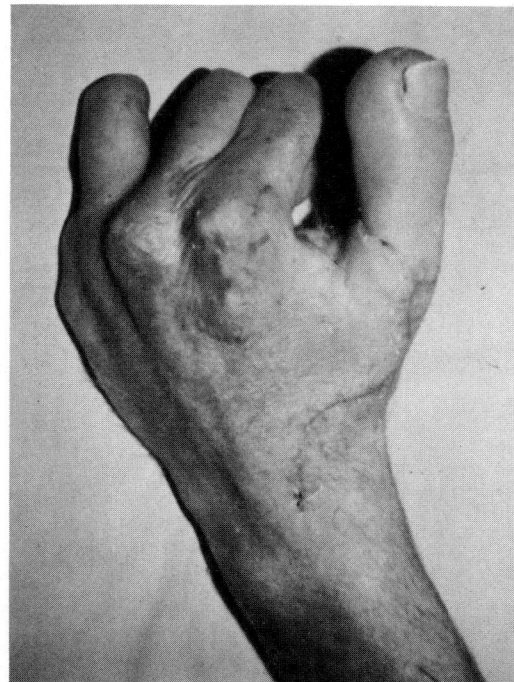

D

22. RECONSTRUCTION OF THE NAIL FOLD

B. Barfod

The dorsal surfaces of the finger and the nail fold often suffer from third degree burns and the destroyed skin can be replaced by partial thickness skin grafts, but these only take on the matrix, resulting in a contracted scar without a nail fold (Fig. 22.1). This scar is ugly, but more importantly the contracture inhibits flexion of the distal interphalangeal joint, which is functionally disabling. To restore flexion to the joint, the scar tissue must be replaced with a skin flap with good elasticity. We have devised a technique for reconstruction of the nail fold, using flaps rotated from the lateral sides of the finger.

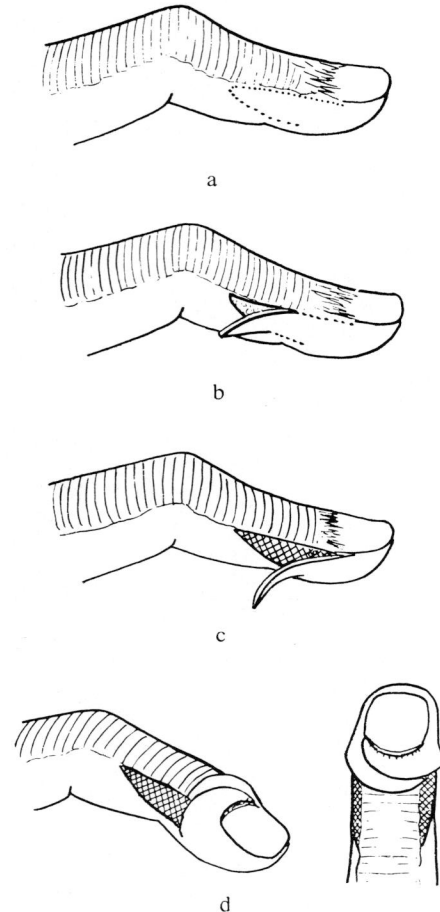

Figure 22.2
Four stages in mobilisation of the skin flaps.

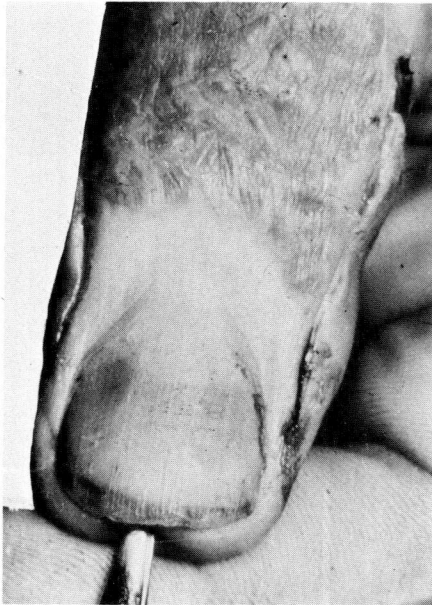

Figure 22.1
Thumb showing contracted scar at the base of the nail. The skin flaps have been marked out.

TECHNIQUE

Flaps must come from a part of the finger where the skin is intact; the palmar surface cannot be used for this purpose. On each side of the finger a narrow triangular flap, based distally, is mobilised. These flaps must be long enough to reach the other side of the finger after rotation. As the length of each flap is three to four times greater than the width, with a distal base, the vascular supply is generally insufficient and the operation must therefore be done in stages.

STAGE 1

The flap is outlined (Fig. 22.2a). Its dorsal margin is placed along the lateral border of the skin graft, which covers the dorsal surface of the finger. The palmar margin must not encroach on the pulp and interfere with its function. The incisions extend through the skin and subcutaneous tissue. The dorsal incision must be carried further distally than the palmar incision to facilitate rotation (Fig. 22.3). The flap is not elevated and the incisions are sutured with 5/0 sutures. The operation is done under a ring block.

STAGE 2

Ten days later the incisions are re-opened and half the flap is mobilised (Fig. 22.2b) and sutured back again in the same place. If the flap is very long and narrow, this procedure may have to be repeated again a third time; on this occasion extending the mobilisation to two-thirds of its length eight to ten days later.

STAGE 3

After ten days have elapsed, the flaps are mobilised avoiding injury to the terminal branches of the digital nerves, which can be seen at the base of the wound. Full or split skin grafts are used to cover the residual defects (Fig. 22.2c), and the flaps are again restored to their original place, covering the free grafts.

STAGE 4

Six days later, the flaps are elevated once more. The scar tissue covering the matrix is resected (Fig. 22.3) and the two strips are rotated towards each other and sutured as shown in Figures 22.2d and 22.3.

The two 'ears' produced by the rotation of the flaps disappear gradually. At best the new nail fold is of normal appearance, but a small notch may remain in its centre.

Figure 22.4 shows the end result of this procedure in three of the fingers of a mechanic, who had sustained burns on the dorsal aspect of the fingers of both hands. Figure 22.1 shows the pre-operative condition in one of his fingers. Following the operation flexion of the terminal interphalangeal joints was restored and he was able to continue his work as a mechanic.

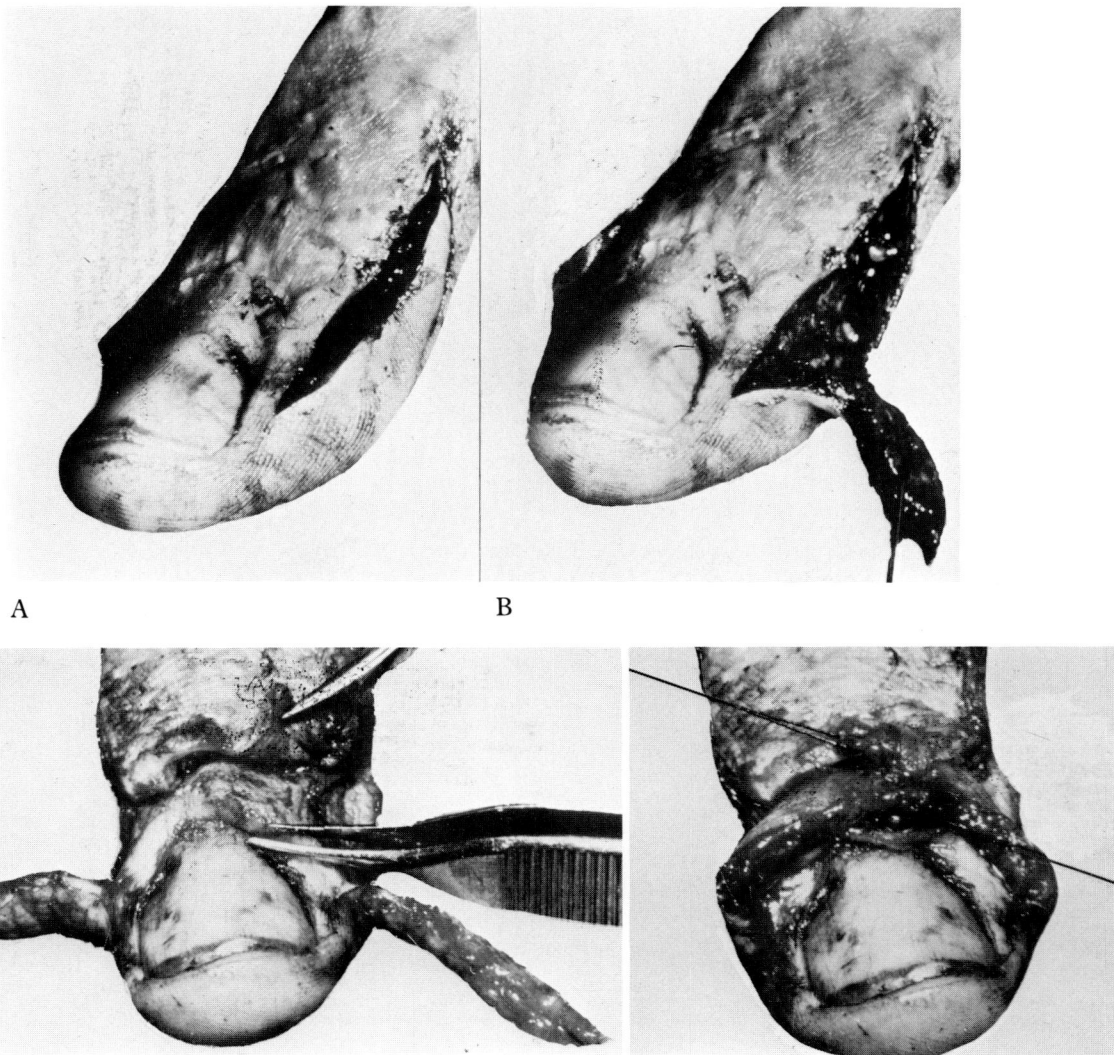

A   B   C   D

Figure 22.3
A and B First and third stages before the skin graft is put in place.
C and D Stages before and after the skin flaps have been rotated into place.

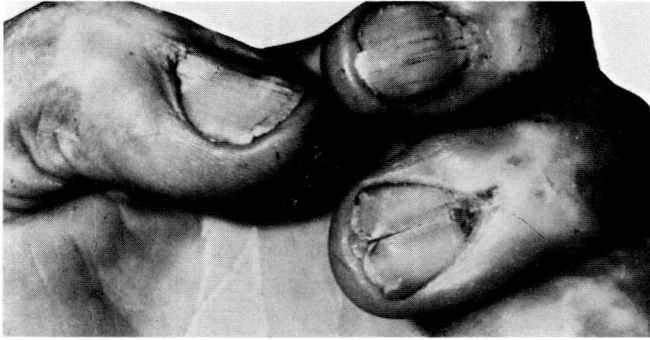

Figure 22.4
Final result.

DISCUSSION

Since this work was first published in 1972 (Hayls, 1972) there appears to have been no further article on this subject. The appearance and function of the finger is much improved following surgery so that the operation appears entirely justified. The fact that the patient, illustrated in Figure 22.4, after undergoing operation on the most severely affected fingers, requested that the other fingers be operated on, demonstrates its value to him and his work.

BIBLIOGRAPHY

[1] Hayls C. W. — One-stage nail fold reconstruction, *The Hand*, 1972, 6, 74–75.

acrokeratosis paraneoplasia, 55
acrosclerosis, 26
Addison's disease, 13
alopecia, 17
  areata, 29, 51–52
  ungium, 24
amniotic disease, 16
anaemia, hypochromic, 29
anonychia, 5, 15, 23
  aplastic, 23
  atrophic, 23
  hyperkeratotic, 23
Apert's syndrome, 17
arsenic poisoning, 29, 33
arthritis, rheumatoid, 26, 29
Aspergillus, 43
atrophy, 23, 28, 29
  idiopathic, 47–48
avulsion injuries, 83–84
  treatment by replacement, 83

Bart's acquired periungual fibrokeratoma, 58
basal cell carcinoma *see* carcinoma, basal cell
Beau's lines, 27–28
biopsy, 2
  longitudinal, 2
  of the periungual fold, 2
  punch, 2
  techniques, 2
Bowen's disease, 54
brachyonychia, 22
bullous epidermolysis, 5

Candida, 40, 43, 44
Candida albicans, 40, 43
Candida tropicalis, 43
  see also onychomycosis, due to *Candida*
carcinoma
  basal cell, 54
  bronchogenic, 61
  squamous cell, 54
cardiopulmonary disorders, 61
causalgia of the median nerve, 26
chromonychia, 30–37, 43
claw nail, 93–101
clubbing, 19–20
  hypertrophic osteoarthropathy, 20
  pachydermoperiostitis, 20
  shell nail syndrome, 20
  simple type, 20
collagen diseases, 61–62
congenital heart disease, 61
Cowden's syndrome, 59
craniocleidodysostosis, 17
Cronkhite-Canada syndrome, 62
Crouzon's disease, 17
crush injuries, 81–82
  complications, 82
  treatment, 81–82
  with fracture of terminal phalanx, 84
  without fracture, 83
cyst, mucous, 26
  see also pseudocyst, mucous

Darier's disease, 48

defluvium, 24
dermatitis, 49
dermatomyositis, 61–62
dermatoses, 6
disease
  acquired, 5, 6–13
  congenital, 5–6
  dermatological, 46–52
  genetic, 65–72
  infectious, 29, 62
  value and limitations of histological
    diagnosis, 3
  see also individual diseases
dyschromia, 30–37
dyskeratoses, congenital, 17
dysplasia, congenital ectodermal, 17
dystrophia mediana canaliform, 26

ectodermis, 15
eczema, 27, 29, 48–49
enchondroma, 57
endocrine disorders, 63
epiloia, 58
epionyx, 15
eponychium, 1, 15
  lesions, 4
escavenitis, 49
exostosis, 57

fibromas, 58
  Steel's garlic clove, 58

gastro-intestinal disorders, 62
genetic disease, *see* disease, genetic
glomus tumours, 57, 76–78
  classification, 77
  clinical features, 76
  clinical types, 77
  examination, 76–77
  histological types, 77
  microscopic structure, 76
  pathological anatomy, 76
  treatment, 77–78
graft
  bone *see* osteo-plastic lengthening
  nail, 104
    two stage toe-to-finger transfer, 104
  skin, 84, 85
    cross finger flap, 96–97
    palmar advancement flap, 96, 97–98
  see also reconstruction, of the nail fold
green nail syndrome, 43–44
grooves, 1, 26
  distal, 1
  lateral, 1
  proximal, 1
  transverse *see* Beau's lines

haemangiopericytoma *see* glomus tumour
Hallopeau's acrodermatitis and
  acropustuloses, 49–50
hapalonychia, 29
herpes, 50
hippocratic nail *see* clubbing

Hjorth-Sabouraud's pustular parakeratosis
  see parakeratosis, Hjorth-Sabouraud's
  pustular
Hodgkin's disease 63
hyperonychia, 21–22
hypertrophic osteoarthropathy *see* clubbing,
  hypertrophic osteoarthropathy
hyponychia, 5
hyponychium, 1
  lesions, 4
ichthyosis, 5, 17
idiopathic atrophy *see* atrophy, idiopathic
incontinentia pigmentia, 59
infectious diseases *see* disease, infectious
ischaemia, vaso-motor, 26

Jadassohn-Lewandowsky's syndrome, 21
junctional naevus of the matrix, 54–55

keratin, 3–4
  injuries to, 29
keratoacanthoma, 57
koilonychia, 5, 20–21
  see also leukokoilonychia

Leclercq's chevron-shaped, median, ungual
  dystrophy, 26
lesions, histological, 4
  see also individual parts of the nail
Letterer-Siwe disease, 63
leukokoilonychia, 5–6
  see also koilonychia *and* leukonychia
leukonychia, 5, 11, 30–37
  acquired forms, 33
  apparent, 30, 33–34
  caused by nail plate involvement, 33
  dermatological forms, 34
  endogenous, 33
  punctate, 33
  striated, 30
  subtotal, 30
  total, 30
  true, 30–33
  see also leukokoilonychia
lichen planus, 26, 29, 46–48
  ulcerative, 47
lichen striatus, 47
lobster-claw hand, 16
lunula, 1–2
lupus erythematosus
  disseminated, 26
  systemic, 61–62

melanin pigment, 11–13
melanoma, 11–13, 74–75
  malignant, 54–55
  subungual, 55, 74
  treatment of, 74–75
microdactyly, 16
microvascular transfer *see* reconstruction, of
  the nail by microvascular transfer
molluscum sebaceum see keratoacanthoma
mongolism, 17
Morey and Burke's nail, 33
Morgagni-Stewart-Morel syndrome, 64

mucinosis, follicular, 63
Muehrcke's paired narrow white bands, 33

naevus striatus symetricus unguis
  see Leclercq's chevron-shaped, median,
    ungual dystrophy
nail
  acquired diseases *see* disease, acquired
  anomalies, 15–18
    in malformations of the hand, 16–17
  arched, 22
  colour, 30–37
  congenital abnormalities, 17
  congenital diseases *see* disease, congenital
  dermatological diseases *see* disease,
    dermatological
  discolouration, 10
  embryology, 15
  examination, 40–42
  fragility, 29
  genetic diseases *see* disease, genetic
  growth, 1
  increased transverse curvature, 22–23
  in the foetus, 3, 15
  layers, 3–4
  malformations, 15–18
  pathological, 5–14
  pincer, 22
  pitting, 28
  shedding *see* onychomadesis
  shortened *see* brachyonychia
  structure, 1
  thickened *see* hyperonychia
  title-shaped, 22–23
  trumpet, 22
  worn, 23
nail atrophy *see* atrophy
nail bed, 1
  lesions, 4
nail matrix, 1
  excision of, 93
  lesions, 4
nail patella syndrome *see*
    osteo-onychodysplasia
nail plate, 1
  see also leukonychia, caused by nail plate
    involvement
nail-biting, 28
nails
  brittle, 29
  friable, 29
  soft, 29
  white *see* leukonychia
nervous system, diseases of, 64
neuromyoarterial glomangioma *see* glomus
    tumour
neurotic nail damage, 28
Norwegian scabies *see* onychia, parasitic

onychatrophy, 15
onychauxis, 21
onychia, 39–44
  bacterial, 42
  mycotic, 39–40
  parasitic, 42
onychodystrophies, 10
onychogryphosis, 21–22
  fungal, 22
onychoheterotopia, 15
onycholysis, 10, 24–25, 43
onychomadesis, 23–24

onychomycosis, 8–10, 40–42
  dermatophytic, 8–9, 42
  distal subungual, 40
  due to *Candida*, 8, 9–10, 40
  proximal white subungual, 40
  superficial white, 29, 40
onychoptosis, 24
onychorrhexis, 26, 29
onychoschizia, lamellar, 29
osteomalacia, 29
osteo-onychodysplasia, 17–18
osteo-plastic lengthening, 94–98
  operative technique, 94–98
osteoporosis, 29

pachydermoperiostitis *see* clubbing,
    pachydermoperiostitis
pachyonychia, 5, 16, 21, 24
  congenita, 21
painful subcutaneous nodule *see* glomus
    tumour
parakeratosis
  Hjorth-Sabouraud's pustular, 50
  of psoriasis, 7
paronychia, 9, 27, 42–44
  parasitic, 44
pemphigus, 50
perithelioma *see* glomus tumour
periungual lesions, 74
petaloid nail, 20
pigmentation, 11–13
plastic surgery, 93–101
poisoning, 33
  see also arsenic poisoning
porphyria, 51
prostheses, 85–89
  creation of nail fold, 85
  in amputations of the fingertips, 90–92
  material used, 90–91
  preparation of nail bed, 85
  removal of matrix, 85
prosthesis
  simple nail, 90
    fixation, 90
    results, 90
  with finger attachment, 90
    results, 90
pseudocyst, mucous, 56, 102–103
  surgical excision, 103
  treatment, 102–103
  see also cyst, mucous
Pseudomonas aeruginosa, 43–44
psoriasis, 6–7, 29, 42, 46–47
  histology, 6–7
  pustular, 7
pterygium, 26, 61
  inversum unguis, 26
  ventral, 26
pulp advancement, 93–94
punctate erosions, 28
pyogenic granuloma, 56

ram's horn nail *see* onychogryphosis
radiodermatitis, 80
Raynaud's disease, 61
recession, 98–101
reconstruction
  of the nail by microvascular transfer,
    108–111

anatomy, 109
long-term results, 111
post-operative management, 111
recipient site, 109–111
secondary defect of the toe, 109
technique, 108–111
  of the nail fold, 114–116
  technique, 114–115
reticulohistiocytosis, multicentric, 63
Reye's benign juvenile fibromatosis, 58
ridges, longitudinal, 26–27
Rosenau's depressions, 28

sarcoidosis, 62
Scopulariopsis, 40
  brevicaulis, 40
shell nail syndrome *see* clubbing, shell nail
    syndrome
skin grafting *see* graft, skin
sportsman's toe, 24
squamous cell carcinoma *see* carcinoma,
    squamous cell
staining, 2
  for the detection of ungual mycosis, 8
  in psoriasis, 6
  techniques, 2
Staphylococci, 43
subungual epidermoid inclusions, 58
syndactyly, 17
syphilis, 50–51
syringomyelia, 64
systemic amyloidosis, 59

tennis toe *see* sportsman's toe
Terry's onychodermal band, 2
Terry's white nail, 33, 62
thimble prosthesis *see* prosthesis, with finger
    attachment
toxaemia, 29
trachyonychia, 28–29
Tranquilli-Leale V–Y advancement, 93, 94, 98
Trichophyton rubrum, 40
  interdigitale, 40
tuberous sclerosis *see* epiloia
tumours, 11–13, 54–59
  benign, 55–59
    in general disease, 58
  carcinomatous *see* carcinoma
  glomus *see* glomus tumours
  Koenen's, 58–59
  malignant, 54–55
  melanotic *see* melanoma
  metastatic, 54
  ungual, 11–13

ungual tumours *see* tumours, ungual
uraemic half and half nail, 34

vascular disease, peripheral, 61
vitamin deficiency, 29
V–Y advancement *see* pulp advancement *and*
    Tranquilli-Leale V–Y advancement

wart, 26
  periungual, 55
  subungual, 55
Wolfe grafting *see* graft, nail

yellow nail syndrome, 62